FEYNMAN
MOTIVES

FEYNMAN
MOTIVES

Matilde Marcolli

California Institute of Technology, USA

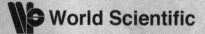

World Scientific

NEW JERSEY · LONDON · SINGAPORE · BEIJING · SHANGHAI · HONG KONG · TAIPEI · CHENNAI

Published by

World Scientific Publishing Co. Pte. Ltd.
5 Toh Tuck Link, Singapore 596224
USA office: 27 Warren Street, Suite 401-402, Hackensack, NJ 07601
UK office: 57 Shelton Street, Covent Garden, London WC2H 9HE

British Library Cataloguing-in-Publication Data
A catalogue record for this book is available from the British Library.

FEYNMAN MOTIVES

ISBN-13 978-981-4271-20-2
ISBN-10 981-4271-20-9
ISBN-13 978-981-4304-48-1 (pbk)
ISBN-10 981-4304-48-4 (pbk)

Printed in Singapore.

Звезда в Галактике и ворон на суку
живут, не мучимы шекспировской дилеммой.
Но мы — мы поздний стиль стареющей Вселенной.
Она сквозь нас поет свою тоску.

Yuri Manin – *"(not yet quite) The Late Style"*

Preface

This book originates from the notes of a course on "Geometry and Arithmetic of Quantum Fields", which I taught at Caltech in the fall of 2008. Having just moved to Caltech and having my first chance to offer a class there, I decided on a topic that would fall in between mathematics and theoretical physics. Though it inevitably feels somewhat strange to be teaching Feynman diagrams at Caltech, I hope that having made the main focus of the lectures the yet largely unexplored relation between quantum field theory and Grothendieck's theory of motives in algebraic geometry may provide a sufficiently different viewpoint on the quantum field theoretic notions to make the resulting combination of topics appealing to mathematicians and physicists alike.

I am not an expert in the theory of motives and this fact is clearly reflected in the way this text is organized. Interested readers will have to look elsewhere for a more informative introduction to the subject (a few references are provided in the text). Also I do not try in any way to give an exhaustive viewpoint of the current status of research on the connection between quantum field theory and motives. Many extremely interesting results are at this point available in the literature. I try, whenever possible, to provide an extensive list of references for the interested reader, and occasionally to summarize some of the available results, but in general I prefer to keep the text as close as possible to the very informal style of the lectures, possibly at the cost of neglecting material that should certainly be included in a more extensive monograph. In particular, the choice of material covered here focuses mostly on those aspects of the subject where I have been actively engaged in recent research work and therefore reflects closely my own bias and personal viewpoint.

In particular, I will try to illustrate the fact that there are two possible complementary approaches to understanding the relation between Feynman integrals and motives, which one may refer to as a "bottom-up" and a "top-down" approach. The bottom-up approach looks at individual Feynman integrals for given Feynman graphs and, using the parametric representation in terms of Schwinger and Feynman parameters, identifies directly the Feynman integral (modulo the important issue of divergences) with an integral of an algebraic differential form on a cycle in an algebraic variety. One then tries to understand the motivic nature of the piece of the relative cohomology of the algebraic variety involved in the computation of the period, trying to identify specific conditions under which it will be a realization of a very special kind of motive, a mixed Tate motive. The other approach, the top-down one, is based on the formal properties that the category of mixed Tate motives satisfies, which are sufficiently rigid to identify it (via the Tannakian formalism) with a category of representations of an affine group scheme. One then approaches the question of the relation to Feynman integrals by showing that the data of Feynman integrals for all graphs and arbitrary scalar field theories also fit together to form a category with the same properties. This second approach was the focus of my joint work with Connes on renormalization and the Riemann–Hilbert correspondence and is already presented in much greater detail in our book "Noncommutative geometry, quantum fields, and motives". However, for the sake of completeness, I have also included a brief and less detailed summary of this aspect of the theory in the present text, referring the readers to the more complete treatment for further information. The bottom-up approach was largely developed in the work of Bloch–Esnault–Kreimer, though in this book I will mostly relate aspects of this approach that come from my joint work with Aluffi. I will also try to illustrate the points of contact between these two different approaches and where possible new developments may arise that might eventually unify the somewhat fragmentary picture we have at the moment. This book only partially reflects the state of the art on this fast-moving subject at the specific time when these lectures were delivered, but I hope that the timeliness of circulating these ideas in the community of mathematicians and physicists will somehow make up for lack of both rigor and of completeness.

Matilde Marcolli

Acknowledgments

A large part of the material covered in this book is based on recent joint work with Paolo Aluffi and on previous joint work with Alain Connes, whom I thank for their essential contributions to the development of those ideas and results. I also thank Abhijnan Rej for conversations and collaboration on some of the topics in this book, and for providing detailed comments on an earlier draft of the manuscript. I thank the students of my "Geometry and Arithmetic of Quantum Fields" course, especially Domenic Denicola and Michael Smith, for their valuable comments and questions during the class. Many thanks are due to Paolo Aluffi for his very detailed and useful feedback on the final version of the manuscript. I also thank Arthur Greenspoon for a final proofreading of the text. This book was partly written during stays of the author at MSRI and at the MPI. I thank both institutes for the hospitality and for support. This work was partially supported by NSF grants DMS-0651925 and DMS-0901221.

Contents

Chapter 1

Perturbative quantum field theory and Feynman diagrams

This first chapter gives a very brief and informal introduction to perturbative quantum field theory. We start with a simple finite dimensional example, which is aimed at illustrating the main ideas in a context familiar to most readers, where the more serious difficulties of the infinite dimensional setting are not present, but one can already see the logic behind the perturbative expansion as well as the roles of Feynman graphs. We then briefly describe the infinite dimensional setting, in the case of a scalar field theory, and the corresponding Feynman rules. We also include a brief discussion of dimensional regularization (DimReg) of divergent Feynman integrals, a subject to which we return later in the book from a more geometric perspective. We include here also a brief discussion of some combinatorial properties of Feynman graphs, which we need in later chapters. We state the renormalization problem, which we discuss in more detail in a later chapter, in the setting of the Connes–Kreimer approach. There are many very good books on quantum field theory. The subject is very rich and can be approached from many different perspectives. For the material we are going to cover in this text, the reader may wish to consult [Bjorken and Drell (1964)], [Bjorken and Drell (1965)], [Itzykson and Zuber (2006)], [LeBellac (1991)], [Nakanishi (1971)]. A lively and very readable introduction, by which the beginning part of this chapter is inspired, can be found in the recent book [Zee (2003)].

1.1 A calculus exercise in Feynman integrals

To understand the role of Feynman graphs in perturbative quantum field theory, it is useful to first see how graphs arise in the more familiar setting of finite dimensional integrals, as a convenient way of parameterizing the terms in the integration by parts of polynomials with respect to a Gaussian measure. It all starts with the simplest Gaussian integral

$$\int_{\mathbb{R}} e^{-\frac{1}{2}ax^2}\, dx = \left(\frac{2\pi}{a}\right)^{1/2}, \tag{1.1}$$

for $a > 0$, which follows from the usual polar coordinates calculation

$$\int_{-\infty}^{\infty} e^{-\frac{1}{2}ax^2}\, dx \int_{-\infty}^{\infty} e^{-\frac{1}{2}ay^2}\, dy = 2\pi \int_0^{\infty} e^{-\frac{1}{2}ar^2}\, r\, dr = \frac{2\pi}{a}\int_0^{\infty} e^{-u}\, du.$$

Similarly, the Gaussian integral with source term is given by

$$\int_{\mathbb{R}} e^{-\frac{1}{2}ax^2+Jx}\, dx = \left(\frac{2\pi}{a}\right)^{1/2} e^{\frac{J^2}{2a}}. \tag{1.2}$$

This also follows easily from (1.1), by completing the square

$$-\frac{ax^2}{2} + Jx = -\frac{a}{2}\left(x^2 - \frac{2Jx}{a}\right) = -\frac{a}{2}\left(x - \frac{J}{a}\right)^2 + \frac{J^2}{2a}$$

and then changing coordinates in the integral to $y = x - \frac{J}{a}$. In this one-dimensional setting a first example of computation of an expectation value can be given in the form

$$\langle x^{2n}\rangle := \frac{\int_{\mathbb{R}} x^{2n}\, e^{-\frac{1}{2}ax^2}\, dx}{\int_{\mathbb{R}} e^{-\frac{1}{2}ax^2}\, dx} = \frac{(2n-1)!!}{a^n}, \tag{1.3}$$

where $(2n-1)!! = (2n-1)\cdot(2n-3)\cdots 5\cdot 3\cdot 1$. One obtains (1.3) inductively from (1.1) by repeatedly applying the operator $-2\frac{d}{da}$ to (1.1). It is worth pointing out that the factor $(2n-1)!!$ has a combinatorial meaning, namely it counts all the different ways of connecting in pairs the $2n$ linear terms x in the monomial $x^{2n} = x\cdot x\cdots x$ in the integral (1.3). In physics one refers to such pairings as Wick contractions. As we discuss below, the analog of the Gaussian integrals in the infinite dimensional setting of quantum field theory will be the free field case, where only the quadratic terms are present in the Lagrangian. The one-dimensional analog of Lagrangians that include interaction terms will be integrals of the form

$$Z(J) = \int_{\mathbb{R}} e^{-\frac{1}{2}ax^2+P(x)+Jx}\, dx, \tag{1.4}$$

where $P(x)$ is a polynomial in x of degree $\deg P \geq 3$. The main idea in such cases, which we'll see applied similarly to the infinite dimensional case, is to treat the additional term $P(x)$ as a perturbation of the original Gaussian integral and expand it out in Taylor series, reducing the problem in this way to a series of terms, each given by the integral of a polynomial under a Gaussian measure. Namely, one writes

$$Z(J) = \int_{\mathbb{R}} \left(\sum_{n=0}^{\infty} \frac{P(x)^n}{n!} \right) e^{-\frac{1}{2}ax^2 + Jx} \, dx. \tag{1.5}$$

The perturbative expansion of the integral (1.4) is defined to be the series

$$\sum_{n=0}^{\infty} \frac{1}{n!} \int_{\mathbb{R}} P(x)^n \, e^{-\frac{1}{2}ax^2 + Jx} \, dx. \tag{1.6}$$

Notice then that, for a monomial x^k, the integral above satisfies

$$\int_{\mathbb{R}} x^k \, e^{-\frac{1}{2}ax^2 + Jx} \, dx = \left(\frac{d}{dJ} \right)^k \int_{\mathbb{R}} e^{-\frac{1}{2}ax^2 + Jx} \, dx. \tag{1.7}$$

Using (1.2), this gives

$$\int_{\mathbb{R}} x^k \, e^{-\frac{1}{2}ax^2 + Jx} \, dx = \left(\frac{2\pi}{a} \right)^{1/2} \left(\frac{d}{dJ} \right)^k e^{\frac{J^2}{2a}}.$$

Thus, in the case where the polynomial $P(x)$ consists of a single term

$$P(x) = \frac{\lambda}{k!} x^k,$$

one can rewrite each term in the perturbative expansion using (1.7), so that one obtains

$$\sum_{n=0}^{\infty} \frac{1}{n!} \int_{\mathbb{R}} \left(\frac{\lambda}{k!} x^k \right)^n e^{-\frac{1}{2}ax^2 + Jx} \, dx =$$

$$\sum_{n=0}^{\infty} \frac{1}{n!} \left(\frac{\lambda}{k!} \left(\frac{d}{dJ} \right)^k \right)^n \int_{\mathbb{R}} e^{-\frac{1}{2}ax^2 + Jx} \, dx.$$

Thus, the perturbative expansion can be written in the form

$$Z(J) = \left(\frac{2\pi}{a} \right)^{1/2} \exp\left(\frac{\lambda}{k!} \left(\frac{d}{dJ} \right)^k \right) \exp\left(\frac{J^2}{2a} \right). \tag{1.8}$$

Two examples of this kind that will reappear frequently in the infinite dimensional version are the cubic case with $P(x) = \frac{g}{6} x^3$ and the quartic case with $P(x) = \frac{\lambda}{4!} x^4$.

   To see then how the combinatorics of graphs can be used as a convenient device to label the terms of different order in λ and J in the perturbative series of $Z(J)$, first observe that the term of order λ^α and J^β in $Z(J)$ is produced by the combination of the term of order α in the Taylor expansion of the exponential $\exp(\frac{\lambda}{k!}(\frac{d}{dJ})^k)$ and the term of order $\beta + k\alpha$ in J in the Taylor expansion of the other exponential $\exp(\frac{J^2}{2a})$ in (1.8). All the resulting terms will be of a similar form, consisting of a combinatorial factor given by a ratio of two products of factorials, a power of J, a power of λ and a power of $2a$ in the denominator. The graphs are introduced as a visual way to keep track of the power counting in these terms, which are associated to the vertices and the internal and external edges of the graph. The combinatorial factor can then also be described in terms of symmetries of the graphs.

Here, as in general in perturbative quantum field theory, one thinks of graphs as being constructed out of a set of vertices and a set of half edges. Each half edge has an end that is connected to a vertex and another end that may pair to another half edge or remain unpaired. An internal edge of the graph consists of a pair of half edges, hence it is an edge in the usual graph theoretic sense, connecting two vertices. An external edge is an unpaired half edge attached to a vertex of the graph. The graphs we consider will not necessarily be connected. It is sometimes convenient to adopt the convention that a graph (or connected component of a graph) consisting of a single line should be thought of as consisting of an internal and two external edges.

The way one assigns graphs to monomials of the form $\frac{\lambda^\alpha J^\beta}{a^\kappa}$ is by the following rules.

- To each factor of λ one associates a vertex of valence equal to the degree of the given monomial $P(x) = \frac{\lambda}{k!}x^k$. This means a total of α vertices, each with k half edges attached.
- To each factor J one associates an external edge.
- The power of a^{-1} is then determined by the resulting number of internal edges obtained by *pairing* all the half edges to form a graph.

Notice that the procedure described here produces not one but a finite collection of graphs associated to a given monomial $\frac{\lambda^\alpha J^\beta}{a^\kappa}$, depending on all the different possible pairings between the half edges. This collection of graphs can in turn be subdivided into isomorphism types, each occurring with a given multiplicity, which corresponds to the number of different pairings that produce equivalent graphs. These combinatorial factors are the *symmetry factors* of graphs. To see more precisely how these factors can be computed, we can introduce the analog, in this 1-dimensional toy model, of the Green functions in quantum field theory. The function $Z(J)$

of (1.4) can be thought of as a generating function for the Green functions

$$Z(J) = \sum_{N=0}^{\infty} \frac{J^N}{N!} \int_{\mathbb{R}} x^N e^{-\frac{1}{2}ax^2 + P(x)} \, dx = Z \cdot \sum_{N=0}^{\infty} J^N G_N, \tag{1.9}$$

where $Z = \int_{\mathbb{R}} e^{-\frac{1}{2}ax^2 + P(x)} \, dx$ and the Green functions are

$$G_N = \frac{\int_{\mathbb{R}} \frac{x^N}{N!} e^{-\frac{1}{2}ax^2 + P(x)} \, dx}{\int_{\mathbb{R}} e^{-\frac{1}{2}ax^2 + P(x)} \, dx}. \tag{1.10}$$

Upon expanding out the interaction term $\exp(P(x))$, with $P(x) = \frac{\lambda}{k!} x^k$ and formally exchanging the sum with the integral, one obtains

$$G_N = \frac{\sum_{n=0}^{\infty} \int \frac{x^N}{N!} \frac{(\lambda x^k)^n}{(k!)^n \, n!} e^{-\frac{1}{2}ax^2} \, dx}{\sum_{n=0}^{\infty} \int \frac{(\lambda x^k)^n}{(k!)^n \, n!} e^{-\frac{1}{2}ax^2} \, dx}. \tag{1.11}$$

Using (1.9), we then see that one way of computing the coefficient of a term in $\frac{\lambda^\alpha J^\beta}{a^\kappa}$ in the asymptotic expansion of $Z(J)$ is to count all the pairings (the Wick contractions) that occur in the integration

$$\int_{\mathbb{R}} x^N x^{kn} e^{-\frac{1}{2}ax^2} \, dx. \tag{1.12}$$

As we have seen in (1.3), these are $(N + kn - 1)!!$ for even $N + kn$, with the odd ones vanishing by symmetry. Taking into account the other coefficients that appear in (1.9) and (1.11), one obtains the factor

$$\frac{(N + kn - 1)!!}{N! \, n! \, (k!)^n}.$$

The meaning of this factor in terms of symmetries of graphs can be explained, by identifying $(N + kn - 1)!!$ with the number of all the possible pairings of half edges, from which one factors out $N!$ permutations of the external edges, $k!$ permutations of the half edges attached to each valence k vertex and $n!$ permutations of the n valence k vertices along with their star of half edges, leaving all the different pairings of half edges. These then correspond to the sum over all the possible topologically distinct graphs obtained via these pairings, each divided by its own symmetry factor. Thus, in terms of graphs, the terms of the asymptotic series in the numerator of (1.10) become of the form

$$\sum_{\Gamma \in \text{graphs}} \frac{\lambda^{\#V(\Gamma)} J^{\#E_{ext}(\Gamma)}}{\#\text{Aut}(\Gamma) \, a^{\#E_{int}(\Gamma)}}. \tag{1.13}$$

Notice also how, when computing the terms of the asymptotic series using either the Taylor series of the exponentials of (1.8) or by first using the expansion in Green functions and then the terms (1.12), one is implicitly using the combinatorial identity

$$(2n - 1)!! = \frac{(2n)!}{2^n n!}.$$

Passing from the 1-dimensional case to a finite dimensional case in many variables is notationally more complicated but conceptually very similar. One replaces an integral of the form

$$\int_{\mathbb{R}^N} e^{-\frac{1}{2}x^t A x + Jx} dx_1 \cdots dx_N = \frac{(2\pi)^{N/2}}{(\det A)^{1/2}} e^{\frac{1}{2} J A^{-1} J^t}, \tag{1.14}$$

where the positive real number $a > 0$ of the 1-dimensional case is now replaced by an $N \times N$ real matrix A with $A^t = A$ and positive eigenvalues. The real number J is here an N-vector, with Jx the inner product. The form of (1.14) is obtained by diagonalizing the matrix and reducing it back to the 1-dimensional case. One can again compute the asymptotic series for the integral

$$\int_{\mathbb{R}^N} e^{-\frac{1}{2}x^t A x + P(x) + Jx} dx_1 \cdots dx_N,$$

where the interaction term here will be a polynomial in the coordinates x_i of x, such as $P(x) = \frac{\lambda}{4!}(\sum_{i=1}^N x_i^4)$. One can use the same method of labeling the terms in the asymptotic series by graphs, where now instead of attaching a factor a^{-1} to the internal edges one finds factors $(A^{-1})_{ij}$ for edges corresponding to a Wick contraction pairing an x_i and an x_j.

The conceptually more difficult step is to adapt this computational procedure for finite dimensional integral to a recipe that is used to make sense of "analogous" computations of functional integrals in quantum field theory.

1.2 From Lagrangian to effective action

In the case of a scalar field theory, one replaces the expression $\frac{1}{2}x^2 + P(x)$ of the one-dimensional toy model we saw in the previous section with a non-linear functional, the Lagrangian density, defined on a configuration space of classical fields. Here we give only a very brief account of the basics of perturbative quantum field theory. A more detailed presentation, aimed at giving a self-contained introduction to mathematicians, can be found in the book [Connes and Marcolli (2008)].

In the scalar case the classical fields are (smooth) functions on a space-time manifold, say $\phi \in \mathcal{C}^{\infty}(\mathbb{R}^D, \mathbb{R})$, and the Lagrangian density is given by an expression of the form

$$\mathcal{L}(\phi) = \frac{1}{2}(\partial\phi)^2 - \frac{m^2}{2}\phi^2 - \mathcal{P}(\phi), \qquad (1.15)$$

where $(\partial\phi)^2 = g^{\mu\nu}\partial_{\mu}\phi\partial_{\nu}\phi$ for $g^{\mu\nu}$ the Lorentzian metric of signature $(1, -1, -1, \ldots, -1)$ on $\mathbb{R}^D$ and a summation over repeated indices understood. The *interaction term* $\mathcal{P}(\phi)$ in the Lagrangian is a polynomial in the field ϕ of degree $\deg \mathcal{P} \geq 3$. Thus, when one talks about a scalar field theory one means the choice of the data of the Lagrangian density and the spacetime dimension D. We can assume for simplicity that $\mathcal{P}(\phi) = \frac{\lambda}{k!}\phi^k$. We will give explicit examples using the special case of the ϕ^3 theory in dimension $D = 6$: while this is not a physically significant example because of the unstable equilibrium point of the potential at $\phi = 0$, it is both sufficiently simple and sufficiently generic with respect to the renormalization properties (*e.g.* non-superrenormalizable), unlike the more physical ϕ^4 in dimension $D = 4$.

To the Lagrangian density one associates a classical action functional

$$S_L(\phi) = \int_{\mathbb{R}^D} \mathcal{L}(\phi)d^D x. \qquad (1.16)$$

The subscript L here stands for the Lorentzian signature of the metric and we'll drop it when we pass to the Euclidean version. This classical action is written as the sum of two terms $S_L(\phi) = S_{free,L}(\phi) + S_{int,L}(\phi)$, where the free field part is

$$S_{free,L}(\phi) = \int_{\mathbb{R}^D} \left(\frac{1}{2}(\partial\phi)^2 - \frac{m^2}{2}\phi^2\right)d^D x$$

and the interaction part is given by

$$S_{int,L}(\phi) = -\int_{\mathbb{R}^D} P(\phi)d^D x.$$

The *probability amplitude* associated to the classical action is the expression

$$e^{i\frac{S_L(\phi)}{\hbar}}, \qquad (1.17)$$

where $\hbar = h/2\pi$ is Planck's constant. In the following we follow the convention of taking units where $\hbar = 1$ so that we do not have to write explicitly the powers of $\hbar$ in the terms of the expansions. An observable of a scalar field theory is a functional on the configuration space of the classical fields, which we write as $\mathcal{O}(\phi)$. The *expectation value* of an observable is defined to be the functional integral

$$\langle \mathcal{O}(\phi) \rangle = \frac{\int \mathcal{O}(\phi) e^{iS_L(\phi)} \, \mathcal{D}[\phi]}{\int e^{iS_L(\phi)} \, \mathcal{D}[\phi]}, \tag{1.18}$$

where the integration is supposed to take place on the configuration space of all classical fields. In particular, one has the N-point Green functions, defined here as

$$G_{N,L}(x_1, \ldots, x_N) = \frac{\int \phi(x_1) \cdots \phi(x_N) \, e^{iS_L(\phi)} \, \mathcal{D}[\phi]}{\int e^{iS_L(\phi)} \, \mathcal{D}[\phi]}, \tag{1.19}$$

for which the generating function is given again by a functional integral with source term

$$\int e^{iS_L(\phi) + \langle J, \phi \rangle} \, \mathcal{D}[\phi], \tag{1.20}$$

where J is a linear functional (a distribution) on the space of classical fields and $\langle J, \phi \rangle = J(\phi)$ is the pairing of the space of fields and its dual. If $J = J(x)$ is itself a smooth function then $\langle J, \phi \rangle = \int_{\mathbb{R}^D} J(x) \phi(x) d^D x$.

Although the notation of (1.18) and (1.19) is suggestive of what the computation of expectation values should be, there are in fact formidable obstacles in trying to make rigorous sense of the functional integral involved. Despite the successes of constructive quantum field theory in several important cases, in general the integral is ill-defined mathematically. This is, in itself, not an obstacle to doing quantum field theory, as long as one regards the expression (1.18) as a shorthand for a corresponding asymptotic expansion, obtained by analogy to the finite dimensional case we have seen previously.

A closer similarity between (1.20) and (1.4) appears when one passes to Euclidean signature by a Wick rotation to imaginary time $t \mapsto it$. This has the effect of switching the signature of the metric to $(1, 1, \ldots, 1)$, after factoring out a minus sign, which turns the probability amplitude into the Euclidean version

$$e^{iS_L(\phi)} \mapsto e^{-S(\phi)}, \tag{1.21}$$

with the Euclidean action

$$S(\phi) = \int_{\mathbb{R}^D} \left(\frac{1}{2} (\partial \phi)^2 + \frac{m^2}{2} \phi^2 + \mathcal{P}(\phi) \right) d^D x. \tag{1.22}$$

Thus, in the Euclidean version we are computing functional integrals of the form

$$G_N(x_1, \ldots, x_N) = \frac{\int \phi(x_1) \cdots \phi(x_N) \, e^{-S(\phi)} \, \mathcal{D}[\phi]}{\int e^{-S(\phi)} \, \mathcal{D}[\phi]}, \tag{1.23}$$

for which the generating function resembles (1.4) in the form

$$Z[J] = \int e^{-\int_{\mathbb{R}^D} \left(\frac{1}{2}(\partial\phi)^2 + \frac{m^2}{2}\phi^2 + \mathcal{P}(\phi) + J(x)\phi(x) \right) d^D x} \, \mathcal{D}[\phi], \qquad (1.24)$$

satisfying

$$\frac{Z[J]}{Z[0]} = \sum_{N=0}^{\infty} \frac{1}{N!} \int J(x_1) \cdots J(x_N) \, G_N(x_1, \ldots, x_N) \, d^D x_1 \cdots d^D x_N, \quad (1.25)$$

for

$$Z[0] = \int e^{-\int_{\mathbb{R}^D} \left(\frac{1}{2}(\partial\phi)^2 + \frac{m^2}{2}\phi^2 + \mathcal{P}(\phi) \right) d^D x} \, \mathcal{D}[\phi]. \qquad (1.26)$$

In order to make sense of this functional integral, one uses an analog of the asymptotic expansion (1.6), where one expands out the exponential of the interaction term $S_{int}(\phi) = \int_{\mathbb{R}^D} \mathcal{P}(x) \, d^D x$ of the Euclidean action and one follows the same formal rules about integration by parts as in the finite dimensional case to label the terms of the expansion by graphs. What is needed in order to write the contribution of a given graph to the asymptotic series is to specify the rules that associate the analogs of the powers of λ, J and a^{-1} to the vertices, external and internal edges of the graph. These are provided by the *Feynman rules* of the theory.

1.3 Feynman rules

By analogy to what we saw in the 1-dimensional model, where one writes the Green functions (1.11) in terms of integrals of the form (1.12), and the latter in terms of sums over graphs as in (1.13), one also writes the Green functions (1.23) in terms of an asymptotic series whose terms are parameterized by graphs,

$$\mathcal{G}_N(p_1, \ldots, p_N) = \sum_{\Gamma} \frac{V(\Gamma, p_1, \ldots, p_N)}{\#\mathrm{Aut}(\Gamma)}, \qquad (1.27)$$

where $\mathcal{G}_N(p_1, \ldots, p_N)$ is the Green function in momentum space, *i.e.* the Fourier transform

$$\mathcal{G}_N(p_1, \ldots, p_N) = \int G_N(x_1, \ldots, x_N) e^{i(p_1 x_1 + \cdots + p_N x_N)} \frac{d^D p_1}{(2\pi)^D} \cdots \frac{d^D p_N}{(2\pi)^D}. \qquad (1.28)$$

The reason for writing the contributions of Feynman integrals in momentum space is that in physics one does not only think of the Feynman graphs as computational devices that do the bookkeeping of terms in integration

by parts of polynomials under a Gaussian measure, but one can think of a
diagram as representing a (part of) a physical process, where certain particles with assigned momenta (external edges) interact (vertices) by creation
and annihilation of virtual particles (internal edges). The momenta flowing
through the graph then represent the physical process. In fact, it is clear
from this point of view that what has physical meaning is not so much
an individual graph but the collection of all graphs with given external
edges and assigned external momenta, and among them the subset of all
those with a fixed number of loops. The latter specifies the order in the
perturbative expansion one is looking at. The terms $V(\Gamma, p_1, \ldots, p_N)$ are
constructed according to the Feynman rules as follows.

- Each internal edge $e \in E_{int}(\Gamma)$ contributes a momentum variable
  $k_e \in \mathbb{R}^D$ so that

$$V(\Gamma, p_1, \ldots, p_N) = \int \mathcal{I}_\Gamma(p_1, \ldots, p_N, k_1, \ldots, k_n) \frac{d^D k_1}{(2\pi)^D} \cdots \frac{d^D k_n}{(2\pi)^D},$$
  (1.29)

  for $n = \#E_{int}(\Gamma)$. The term $\mathcal{I}_\Gamma(p_1, \ldots, p_N, k_1, \ldots, k_n)$ is constructed according to the following procedure.

- Each vertex $v \in V(\Gamma)$ contributes a factor of $\lambda_v (2\pi)^D$, where λ_v is
  the coupling constant of the monomial in the Lagrangian of order
  equal to the valence of v and a conservation law for all the momenta
  that flow through that vertex,

$$\delta_v(k) := \delta\left(\sum_{s(e)=v} k_e - \sum_{t(e)=v} k_e \right),$$
  (1.30)

written after chosing an orientation of the edges of the graph. In
the case of vertices with both internal and external edges (1.30) is
equivalently written in the form

$$\delta_v(k, p) := \delta\left(\sum_{i=1}^n \epsilon_{v,i} k_i + \sum_{j=1}^N \epsilon_{v,j} p_j \right),$$
  (1.31)

where the incidence matrix ϵ of the graph Γ is the $\#V(\Gamma) \times \#E(\Gamma)$-
matrix with

$$\epsilon_{v,e} = \begin{cases} +1 & \text{for } v = t(e) \\ -1 & \text{for } v = s(e) \\ 0 & \text{for } v \notin \partial(e). \end{cases}$$
  (1.32)

- Each internal edge contributes an inverse propagator, that is, a term of the form q_e^{-1}, where q_e is a quadratic form, which in the case of a scalar field in Euclidean signature is given by

$$q_e(k_e) = k_e^2 + m^2. \tag{1.33}$$

- Each external edge $e \in E_{ext}(\Gamma)$ contributes a propagator $q_e(p_e)^{-1}$, with $q_e(p_e) = p_e^2 + m^2$. The external momenta are assigned so that they satisfy the conservation law $\sum_e p_e = 0$, when summed over the oriented external edges.

- The integrand $\mathcal{I}_\Gamma(p_1, \ldots, p_N, k_1, \ldots, k_n)$ is then a product

$$\prod_{v \in V(\Gamma)} \lambda_v (2\pi)^D \delta_v(k_{e_i}, p_{e_j}) \prod_{e_i \in E_{int}(\Gamma)} q_{e_i}(k_{e_i})^{-1} \prod_{e_j \in E_{ext}(\Gamma)} q_{e_j}(p_{e_j})^{-1}, \tag{1.34}$$

with linear relations among the momentum variables k_{e_i} and p_{e_j} imposed by the conservation laws $\delta_v(k_{e_i}, p_{e_j})$ at the vertices of the graph.

We can then write the Feynman integral associated to a Feynman graph Γ of the given theory in the form

$$V(\Gamma, p_1, \ldots, p_N) = \varepsilon(p_1, \ldots, p_N) U(\Gamma, p_1, \ldots, p_N), \tag{1.35}$$

where the factor $\varepsilon(p)$ is the product of the inverse propagators of the external edges

$$\varepsilon(p_1, \ldots, p_N) = \prod_{e \in E_{ext}(\Gamma)} q_e(p_e)^{-1}, \tag{1.36}$$

while the factor $U(\Gamma, p)$ is given by

$$U(\Gamma, p_1, \ldots, p_N) = C \int \frac{\delta(\sum_{i=1}^n \epsilon_{v,i} k_i + \sum_{j=1}^N \epsilon_{v,j} p_j)}{q_1(k_1) \cdots q_n(k_n)} \frac{d^D k_1}{(2\pi)^D} \cdots \frac{d^D k_n}{(2\pi)^D}, \tag{1.37}$$

with $C = \prod_{v \in V(\Gamma)} \lambda_v (2\pi)^D$.

We work here for convenience always with Euclidean signature in the Feynman integrals, so that our propagators contain quadratic terms of the form (1.33). The study of parametric Feynman integral we discuss in greater detail in Chapter 3 can be done also for the Lorentzian case, as shown for instance in [Bjorken and Drell (1965)].

1.4 Simplifying graphs: vacuum bubbles, connected graphs

There are some useful simplifications that can be done in the combinatorics of graphs that appear in the formal series (1.27).

The basic property that makes these simplifications possible is the multiplicative form (1.34) of the Feynman integrand $\mathcal{I}_\Gamma(p_1, \ldots, p_N, k_1, \ldots, k_n)$. This implies the following property.

Lemma 1.4.1. *The Feynman integral $V(\Gamma, p_1, \ldots, p_N)$ is multiplicative on connected components of the graph Γ.*

Proof. This follows immediately from the form (1.34) and (1.35) with (1.36), (1.37) of the Feynman integral. In fact, if the graph Γ has different connected components, no linear relations arise between momentum variables of the edges of different components (as these have no common vertices) and the corresponding integrals split as a product. □

Moreover, one also has the following multiplicative form of the symmetry factors of graphs.

Lemma 1.4.2. *For a graph Γ that is a union of connected components Γ_j with multiplicities n_j (i.e. there are n_j connected components of Γ all isomorphic to the same graph Γ_j), the symmetry factor splits multiplicatively on components according to the formula*

$$\#\mathrm{Aut}(\Gamma) = \prod_j (n_j)! \prod_j \#\mathrm{Aut}(\Gamma_j)^{n_j}. \tag{1.38}$$

Proof. The factorials come from the symmetries of the graph Γ that permute topologically equivalent components. All symmetries of Γ are obtained by composing this type of symmetries with symmetries of each component. □

One then has a first useful observation on the combinatorics of the graphs that appear in the asymptotic expansion of the Green functions. A graph with no external edges is commonly referred to as a *vacuum bubble*.

Lemma 1.4.3. *The graphs of (1.27) do not contain any vacuum bubbles.*

Proof. As we have seen in the finite dimensional toy model, these correspond to the terms with $J = 0$ in the asymptotic series. Thus, when one writes the expansion (1.25) into Green functions, and then the expansion (1.27) of the latter into Feynman integrals of graphs, the expansion of the functional integral $Z[J]$ would count the contribution of all graphs including components that are vacuum bubbles as in the case of (1.8) in the finite dimensional case. The expansion of $Z[0]$ on the other hand only has contributions from the vacuum bubble graphs, and the multiplicative properties of Lemma 1.4.1 and Lemma 1.4.2 then imply that the expansion for $Z[J]/Z[0]$ only has contributions from graphs which do not contain any connected components that are vacuum bubbles. $\qquad\square$

One can then pass from multi-connected to connected graphs by rewriting the functional integral $Z[J]$ in an equivalent form in terms of

$$W[J] = \log\left(\frac{Z[J]}{Z[0]}\right). \tag{1.39}$$

One can again write a formal asymptotic series for $W[J]$ as

$$W[J] = \sum_{N=0}^{\infty} \frac{1}{N!} \int J(x_1)\cdots J(x_N)\, G_{N,c}(x_1,\ldots,x_N)\, d^D x_1 \cdots d^D x_N, \tag{1.40}$$

where now the Green functions $G_{N,c}(x_1,\ldots,x_N)$ will also have an expansion on graphs of the form (1.27), where, however, only a smaller class of graphs will be involved.

Lemma 1.4.4. *The* connected *Green functions $G_{N,c}(x_1,\ldots,x_N)$ of (1.40) have an expansion*

$$\mathcal{G}_{N,c}(p_1,\ldots,p_N) = \sum_{\Gamma \text{ connected}} \frac{V(\Gamma,p_1,\ldots,p_N)}{\#\mathrm{Aut}(\Gamma)}, \tag{1.41}$$

where $\mathcal{G}_{N,c}(p_1,\ldots,p_N)$ is the Fourier transform

$$\mathcal{G}_{N,c}(p_1,\ldots,p_N) = \int G_{N,c}(x_1,\ldots,x_N) e^{i(p_1 x_1 + \cdots + p_N x_N)} \frac{d^D p_1}{(2\pi)^D} \cdots \frac{d^D p_N}{(2\pi)^D} \tag{1.42}$$

and the $V(\Gamma,p_1,\ldots,p_N)$ in (1.41) are computed as in (1.35).

Proof. We only sketch briefly why the result holds. More detailed expositions can be found in standard Quantum Field Theory textbooks (for example in [LeBellac (1991)]). Suppose that Γ is a disjoint union of connected components $\Gamma = \cup_j \Gamma_j$, with multiplicities n_j and with N_j external edges, so that $\sum_j n_j N_j = N$. Then by Lemma 1.4.1 and 1.4.2 we get

$$\frac{V(\Gamma,p_1,\ldots,p_N)}{\#\mathrm{Aut}(\Gamma)} = \prod_j \frac{V(\Gamma_j,p_1,\ldots,p_{N_j})^{n_j}}{(n_j)!\,\#\mathrm{Aut}(\Gamma_j)^{n_j}}.$$

Thus, we can write

$$\frac{Z[J]}{Z[0]} = \sum_N \frac{1}{N!} \int J(x_1) \cdots J(x_N) \, G_N(x_1, \ldots, x_N) \prod_i d^D x_i$$

$$= \sum_N \sum_{\sum_j n_j N_j = N} \prod_j \frac{1}{n_j} \left(\int J(x_1) \cdots J(x_{N_j}) G_{N_j,c}(x_1, \ldots, x_{N_j}) \right)^{n_j}$$

$$= \exp \left(\sum_N \frac{1}{N!} \int J(x_1) \cdots J(x_N) \, G_{N,c}(x_1, \ldots, x_N) \right). \qquad \square$$

1.5 One-particle-irreducible graphs

Further simplifications of the combinatorics of graphs can be obtained by passing to the *1PI effective action*, or higher loop versions like the *2PI effective action*, etc. We discuss here briefly only the 1PI effective action, though we will later need to return to discussing higher connectivity conditions on Feynman graphs. We first recall the following notions of connectivity of graphs.

Definition 1.5.1. The notion of k-connectivity of graphs is given as follows:

- A graph is k-edge-connected if it cannot be disconnected by removal of any set of k or fewer edges.
- A graph is 2-vertex-connected if it has no looping edges, it has at least 3 vertices, and it cannot be disconnected by removal of a single vertex, where vertex removal is defined as below.
- For $k \geq 3$, a graph is k-vertex-connected if it has no looping edges and no multiple edges, it has at least $k + 1$ vertices, and it cannot be disconnected by removal of any set of $k - 1$ vertices.

Here what one means by removing a vertex from a graph is the following. Given a graph Γ and a vertex $v \in V(\Gamma)$, the graph $\Gamma \smallsetminus v$ is the graph with vertex set $V(\Gamma) \smallsetminus \{v\}$ and edge set $E(\Gamma) \smallsetminus \{e : v \in \partial(e)\}$, *i.e.* the graph obtained by removing from Γ the star of the vertex v. Thus, 1-vertex-connected and 1-edge-connected simply mean connected, while for $k \geq 2$ the condition of being k-vertex-connected is stronger than that of being k-edge-connected. The terminology more commonly in use in the physics literature is the following.

Definition 1.5.2. For $k \geq 2$ a $(k+1)$-edge-connected graph is also called k-particle-irreducible (kPI). For $k = 1$, a 2-edge-connected graph that is not a tree is called one-particle-irreducible (1PI) graph. These cannot be disconnected by removal of a single (internal) edge.

Notice that trees are all considered *not* to be 1PI, even though a tree consisting of just n edges attached to a single valence n vertex cannot be disconnected by removal of a single edge (such edges are not internal though in this case).

Lemma 1.5.3. *Any connected graph can be obtained from a tree, after replacing the vertices by 1PI graphs with the number of external edges equal to the valence of the vertex.*

Proof. If the connected graph Γ is 1PI the tree consists of a single vertex with the number of edges attached equal to the number of external edges of the graph. Suppose the graph is not 1PI. Find an edge that disconnects the graph. Look at each component and again repeat the operations finding edges that disconnect them further until one is left with a collection of 1PI graphs, $\Gamma_1, \ldots, \Gamma_n$. It then suffices to show that the graph obtained from Γ by shrinking each Γ_i to a vertex is a tree. Since each of the internal edges that remain in this graph was an edge whose removal increased the number of connected components, this must still be true in the graph obtained after collapsing all the Γ_i. A graph that is disconnected by the removal of any one internal edge is a tree with at least one internal edge. $\square$

This suggests that there should be a way to further simplify the combinatorics of graphs in the asymptotic expansion of the functional integrals, by counting separately the contributions of trees and that of 1PI graphs, and getting back from these the contributions of all connected graphs.

This is done by passing to the *1PI effective action*, which is defined as the Legendre transform of the functional $W[J]$, namely

$$S_{\text{eff}}(\phi) = (\langle \phi, J \rangle - W[J])|_{J=J(\phi)}, \tag{1.43}$$

evaluated at a stationary J, that is, a solution of the variational equation

$$\frac{\delta}{\delta J} (\langle \phi, J \rangle - W[J]) = 0.$$

The asymptotic expansion of the effective action collects the contributions of all the 1PI graphs, so that the semiclassical calculations (*i.e.* involving only graphs that are trees) done with the effective action recover the full quantum corrections to the classical action given by all the connected graphs that appear in the expansion of $W[J]$. We do not prove here how one gets the asymptotic expansion for the effective action and we refer the interested reader to [Connes and Marcolli (2008)] and [LeBellac (1991)].

To explain more precisely the difference between edge and vertex connectivity in Definition 1.5.1 above, and their relation to the 1PI condition, we recall the following observations from [Aluffi and Marcolli (2009a)].

Definition 1.5.4. A graph Γ' is a splitting of Γ at a vertex $v \in V(\Gamma)$ if it is obtained by partitioning the set $E \subset E(\Gamma)$ of edges adjacent to v into two disjoint non-empty subsets, $E = E_1 \cup E_2$ and inserting a new edge e to whose end vertices v_1 and v_2 the edges in the two sets E_1 and E_2 are respectively attached, as in the example in the figure.

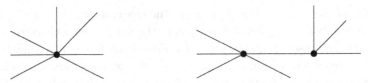

We have the following relation between 2-vertex-connectivity and 2-edge-connectivity (1PI).

Lemma 1.5.5. *Let Γ be a graph with at least 3 vertices and no looping edges.*

(1) If Γ is 2-vertex-connected then it is also 2-edge-connected (1PI).
(2) Γ is 2-vertex-connected if and only if all the graphs Γ' obtained as splittings of Γ at any $v \in V(\Gamma)$ are 2-edge-connected (1PI).

Proof. (1) We have to show that, for a graph Γ with at least 3 vertices and no looping edges, 2-vertex-connectivity implies 2-edge-connectivity. Assume that Γ is not 1PI. Then there exists an edge e such that $\Gamma \smallsetminus e$ has two connected components Γ_1 and Γ_2. Since Γ has no looping edges, e has two distinct endpoints v_1 and v_2, which belong to the two different components after the edge removal. Since Γ has at least 3 vertices, at least one of the two components contains at least two vertices. Assume then that there exists $v \neq v_1 \in V(\Gamma_1)$. Then, after the removal of the vertex v_1 from Γ, the vertices v and v_2 belong to different connected components, so that Γ is not 2-vertex-connected.

(2) We need to show that 2-vertex-connectivity is equivalent to all splittings Γ' being 1PI. Suppose first that Γ is not 2-vertex-connected. Since

Γ has at least 3 vertices and no looping edges, the failure of 2-vertex-connectivity means that there exists a vertex v whose removal disconnects the graph. Let $V \subset V(\Gamma)$ be the set of vertices other than v that are endpoints of the edges adjacent to v. This set is a union $V = V_1 \cup V_2$ where the vertices in the two subsets V_i are contained in at least two different connected components of $\Gamma \smallsetminus v$. Then the splitting Γ' of Γ at v obtained by inserting an edge e such that the endpoints v_1 and v_2 are connected by edges, respectively, to the vertices in V_1 and V_2 is not 1PI.

Conversely, assume that there exists a splitting Γ' of Γ at a vertex v that is not 1PI. There exists an edge e of Γ' whose removal disconnects the graph. If e already belonged to Γ, then Γ would not be 1PI (and hence not 2-vertex connected, by (1)), as removal of e would disconnect it. So e must be the edge added in the splitting of Γ at the vertex v.

Let v_1 and v_2 be the endpoints of e. None of the other edges adjacent to v_1 or v_2 is a looping edge, by hypothesis; therefore there exist at least another vertex $v_1' \neq v_2$ adjacent to v_1, and a vertex $v_2' \neq v_1$ adjacent to v_2. Since $\Gamma' \smallsetminus e$ is disconnected, v_1' and v_2' are in distinct connected components of $\Gamma' \smallsetminus e$. Since v_1' and v_2' are in $\Gamma \smallsetminus v$, and $\Gamma \smallsetminus v$ is contained in $\Gamma' \smallsetminus e$, it follows that removing v from Γ would also disconnect the graph. Thus Γ is not 2-vertex-connected. $\qquad\square$

Lemma 1.5.6. *Let Γ be a graph with at least 4 vertices, with no looping edges and no multiple edges. Then 3-vertex-connectivity implies 3-edge-connectivity.*

Proof. We argue by contradiction. Assume that Γ is 3-vertex-connected but not 2PI. We know it is 1PI because of the previous lemma. Thus, there exist two edges e_1 and e_2 such that the removal of both edges is needed to disconnect the graph. Since we are assuming that Γ has no multiple or looping edges, the two edges have at most one endpoint in common.

Suppose first that they have a common endpoint v. Let v_1 and v_2 denote the remaining two endpoints, $v_i \in \partial e_i$, $v_1 \neq v_2$. If the vertices v_1 and v_2 belong to different connected components after removing e_1 and e_2, then the removal of the vertex v disconnects the graph, so that Γ is not 3-vertex-connected (in fact not even 2-vertex-connected). If v_1 and v_2 belong to the same connected component, then v must be in a different component. Since the graph has at least 4 vertices and no multiple or looping edges, there exists at least one other edge attached to either v_1, v_2, or v, with the other endpoint $w \notin \{v, v_1, v_2\}$. If w is adjacent to v, then removing v and v_1 leaves v_2 and w in different connected components. Similarly, if w is adjacent to (say) v_1, then the removal of the two vertices v_1 and v_2 leave v and w in two different connected components. Hence Γ is not 3-vertex-connected.

Next, suppose that e_1 and e_2 have no endpoint in common. Let v_1 and w_1 be the endpoints of e_1 and v_2 and w_2 be the endpoints of e_2. At least one pair $\{v_i, w_i\}$ belongs to two separate components after the removal of the two edges, though not all four points can belong to different connected components, else the graph would not be 1PI. Suppose then that v_1 and w_1 are in different components. It also cannot happen that v_2 and w_2 belong to the same component, else the removal of e_1 alone would disconnect the graph. We can then assume that, say, v_2 belongs to the same component as v_1 while w_2 belongs to a different component (which may or may not be the same as that of w_1). Then the removal of v_1 and w_2 leaves v_2 and w_1 in two different components so that the graph is not 3-vertex-connected.□

Conditions of 3-connectivity (3-vertex-connected or 3-edge-connected) arise in a more subtle manner in the theory of Feynman integrals, in the analysis of Laundau singularities (see for instance [Sato, Miwa, Jimbo, Oshima (1976)]). In particular, the 2PI effective action is often considered in quantum field theory in relation to non-equilibrium phenomena, see *e.g.* [Rammer (2007)], §10.5.1.

In the following we restrict our attention to Feynman integrals of graphs that are at least 1PI.

1.6 The problem of renormalization

So far we have treated the integrals $U(\Gamma, p_1, \ldots, p_N)$ of (1.37) as purely formal expressions. However, if one tries to assign to such integrals a numerical value, one soon realizes that most of them are in fact divergent. This was historically one of the main problems in the development of perturbative quantum field theory, namely the *renormalization problem*: how to extract in a consistent and physically significant manner finite values from the divergent integrals (1.37) that appear in the asymptotic expansion of the functional integrals of quantum field theory.

The problem of renormalization consists of three main aspects:

- Regularization
- Subtraction
- Renormalization

Regularization consists of a procedure that replaces the divergent integrals (1.37) by functions of some regularization parameters, in such a way that the resulting function has a pole or a divergence for particular values or limits of the additional parameters that correspond to the original divergent integral, but has finite values for other values of the regularization parameters. Subtraction then consists of removing the divergent part of the regularized integrals by a uniform procedure (such as removing the polar part of a Laurent series in the main example we use below). This is not all there is yet. Renormalization means being able to perform the subtraction procedure by modifying the parameters in the Lagrangian (which become themselves functions of the regularization parameter).

To understand more clearly the last point, it is important to stress the fact that the parameters that appear in the Lagrangian, such as masses and coupling constants of the interaction terms, are not physical observables, nor are they the same as the actual masses and physical quantities that one can measure in experiments. In fact, the parameters in the Lagrangian can be modified without affecting the physics one observes. This is what makes it possible, in a renormalizable theory, to correct for divergent graphs by readjusting the parameters in the Lagrangian. Thus, if one introduces a regularization of the divergent Feynman integrals in terms of a complex parameter z (as we discuss below) or in terms of a cutoff Λ, then the Lagrangian can be modified by changing the coefficients to

$$\mathcal{L}(\phi) = \left(\frac{1 + \delta Z}{2} (\partial \phi)^2 + \frac{m^2 + \delta m^2}{2} \phi^2 + \frac{\lambda + \delta \lambda}{k!} \phi^k \right), \qquad (1.44)$$

where the functions δZ, δm^2 and $\delta \lambda$ depend on the regularization parameter. They consist, in fact, of a formal series of contributions coming from all the divergent graphs of the theory. Notice that a theory is still renormalizable if, in addition to the modifications of the coefficients of the terms initially present in the Lagrangian, to compensate for divergent graphs one needs to also add a *finite* number of other terms, *i.e.* other monomials $\frac{\delta \lambda_i}{(k_i)!} \phi^{k_i}$, that were not initially present. In fact, one can just think of this as being the effect of having chosen an arbitrary value $= 0$ for the coefficients of these terms in the initial Lagrangian. However, a theory is no longer renormalizable if an infinite number of additional terms is needed to compensate for the divergences.

1.7　Gamma functions, Schwinger and Feynman parameters

We digress momentarily to recall some useful formulae involving Gamma functions, which are extensively used in Feynman integral computations and are the basis of both the dimensional regularization procedure we describe below and the parametric representation of Feynman integrals that we discuss later and which is the basis for the relation between Feynman integrals and periods of algebraic varieties.

First recall that the Gamma function is defined by the integral

$$\Gamma(t+1) = \int_0^\infty s^t e^{-s} ds. \tag{1.45}$$

It satisfies $\Gamma(t+1) = t\,\Gamma(t)$, hence it extends the factorial, namely $\Gamma(n+1) = n!$ for nonnegative integers. The function $\Gamma(t)$ defined in this way extends to a meromorphic function with poles at all the non-positive integers.

As we are going to see in more detail below, a typical way of dealing with divergences in Feynman integrals is to first identify them with poles of some meromorphic function and typically a product of Gamma functions.

The first useful operation on Feynman integrals is the introduction of *Schwinger parameters*. These are based on the very simple identity

$$\frac{1}{q} = \int_0^\infty e^{-sq} ds. \tag{1.46}$$

This allows the reformulation of integrals where quadratic forms q_i in the momentum variables appear in the denominator in terms of Gaussian integrals where quadratic forms appear in the exponent. In terms of Schwinger parameters, one writes the denominator of a Feynman integral of the form (1.37) as

$$\frac{1}{q_1 \cdots q_n} = \int_{\mathbb{R}_+^n} e^{-(s_1 q_1 + \cdots + s_n q_n)} ds_1 \cdots ds_n. \tag{1.47}$$

This is a special case of the more general useful identity

$$\frac{1}{q_1^{k_1} \cdots q_n^{k_n}} = \frac{1}{\Gamma(k_1) \cdots \Gamma(k_n)} \int_{\mathbb{R}_+^n} e^{-(s_1 q_1 + \cdots + s_n q_n)} s_1^{k_1-1} \cdots s_n^{k_n-1} ds_1 \cdots ds_n. \tag{1.48}$$

Another related way to reformulate the expression (1.37) of Feynman integrals is based on the *Feynman parameters*. Here the basic example, analogous to (1.46) is the *Feynman trick*

$$\frac{1}{ab} = \int_0^1 \frac{1}{(ta + (1-t)b)^2} dt. \tag{1.49}$$

The more general expression analogous to (1.47) is obtained by considering the general formula (1.48) and performing a change of variables $s_i = St_i$ with $S = s_1 + \cdots + s_n$ so as to obtain

$$\frac{1}{q_1^{k_1} \cdots q_n^{k_n}} = \frac{\Gamma(k_1 + \cdots + k_n)}{\Gamma(k_1) \cdots \Gamma(k_n)} \int_{[0,1]^n} \frac{t_1^{k_1-1} \cdots t_n^{k_n-1} \, \delta(1 - \sum_i t_i)}{(t_1 q_1 + \cdots + t_n q_n)^n} \, dt_1 \cdots dt_n.$$
(1.50)

In the particular case of the denominators of (1.37) this gives

$$\frac{1}{q_1 \cdots q_n} = (n-1)! \int_{[0,1]^n} \frac{\delta(1 - \sum_i t_i)}{(t_1 q_1 + \cdots + t_n q_n)^n} \, dt_1 \cdots dt_n. \qquad (1.51)$$

Thus, the integration is performed on the n-dimensional *topological simplex*

$$\sigma_n = \{t = (t_1, \ldots, t_n) \in \mathbb{R}_+^n \mid \sum_i t_i = 1\}. \qquad (1.52)$$

1.8 Dimensional Regularization and Minimal Subtraction

Dimensional regularization (DimReg) is based on a formal procedure aimed at making sense of integrals in "complexified dimension" $D-z$, with $z \in \mathbb{C}^*$, instead of integral dimension $D \in \mathbb{N}$. It would seem at first that one needs to develop a notion of geometry in "complexified dimension" along with measure spaces and a suitable theory of integration, in order to define dimensional regularization property. In fact, much less is needed. Due to the very special form of the Feynman integrals $U(\Gamma, p_1, \ldots, p_N)$, it suffices to have a good definition for the Gaussian integrals in D dimensions,

$$\int e^{-\lambda t^2} d^D t = \pi^{D/2} \lambda^{-D/2}, \qquad (1.53)$$

in the case where D is no longer a positive integer but a complex number. In fact, one can then reformulate the Feynman integrals in terms of Gaussian integrals, using the method of Schwinger parameters described briefly in §1.7.

Clearly, since the right hand side of (1.53) continues to make sense for $D \in \mathbb{C}$, one can use the right hand side of (1.53) as the *definition* of the left hand side and set

$$\int e^{-\lambda t^2} d^z t := \pi^{z/2} \lambda^{-z/2}, \qquad \forall z \in \mathbb{C}. \qquad (1.54)$$

One then obtains well defined Feynman integrals in complexified dimension $D - z$, which one writes formally as

$$U_\mu^z(\Gamma(p_1, \ldots, p_N)) = \int \mathcal{I}_\Gamma(p_1, \ldots, p_N, k_1, \ldots, k_\ell) \, \mu^{z\ell} \, d^{D-z}k_1 \cdots d^{D-z}k_\ell.$$

$$(1.55)$$

The variable μ has the physical units of a mass and appears in these integrals for dimensional reasons. It will play an important role later on, as it sets the dependence on the energy scale of the renormalized values of the Feynman integrals, hence the renormalization group flow.

It is not an easy result to show that the dimensionally regularized integrals give meromorphic functions in the variable z, with a Laurent series expansion at $z = 0$. See a detailed discussion of this point in Chapter 1 of [Connes and Marcolli (2008)]. We will not enter into details here and talk loosely about (1.55) as a meromorphic function of z depending on the additional parameter μ.

The method of *minimal subtraction* consists of removing the divergent part of the Laurent series expansion at $z = 0$ of a meromorphic function obtained from the regularization of the Feynman integrals. One denotes by $\mathfrak{T}$ the operator of projection of a Laurent series onto its polar part. With this notation, if $U(\Gamma) = U(\Gamma)(z)$ is shorthand for the meromorphic function obtained from the dimensional regularization of the Feynman integral associated to the graph Γ (suppressing the explicit dependence on the external momenta), then $(1 - \mathfrak{T})U(\Gamma)$ would be its minimal subtraction. This is a convergent power series in the variable $z \in \mathbb{C}^*$, which then has a finite value at $z = 0$.

As we see below (see also [Connes and Marcolli (2008)] for a more detailed explanation), taking $(1 - \mathfrak{T})U(\Gamma)|_{z=0}$ does not suffice as a renormalization method, due to the role of subdivergences, *i.e.* smaller subgraphs γ of the Feynman graph Γ which themselves already contribute divergent Feynman integrals $U(\gamma)$. The correct procedure that uses dimensional regularization and minimal subtraction to extract finite values from divergent Feynman integrals is given by the Bogolyubov recursion described below in §5.1.

We report here from [Connes and Marcolli (2008)], [Collins (1986)] the simplest example of dimensional regularization, for the self-energy graph of the theory $\mathcal{T}$ with $\mathcal{L}_f(\phi) = \phi^3$ and $D = 6$.

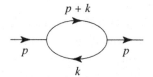

This corresponds to the divergent Feynman integral (neglecting the constant multiplicative factor containing the coupling constant λ and powers of 2π) is given by

$$\int \frac{1}{k^2 + m^2} \frac{1}{(p+k)^2 + m^2} d^D k$$

for the theory with Lagrangian $\mathcal{L}(\phi) = \frac{1}{2}(\partial\phi)^2 + \frac{m^2}{2}\phi^2 + \frac{\lambda}{6}\phi^3$. This follows directly from the application of the Feynman rules: the variables assigned to the two internal edges are $-k$ and $p+k$, with p the external momentum, due to the effect of the delta functions imposing the conservation law at the two vertices. It is easy to see that this integral is divergent, for instance in dimension $D = 4$ or $D = 6$, where it is usually interesting to consider this scalar field theory.

In this case, the method of *Schwinger parameters* described in §1.7 consists of replacing the integral above by

$$\frac{1}{k^2 + m^2} \frac{1}{(p+k)^2 + m^2} = \int_{s>0,\, t>0} e^{-s(k^2+m^2)-t((p+k)^2+m^2)} \, ds \, dt.$$

One can then diagonalize the quadratic form in the exponential to get

$$-Q(k) = -\lambda \left((k + xp)^2 + ((x - x^2)p^2 + m^2) \right)$$

with $s = (1 - x)\lambda$ and $t = x\lambda$. One obtains in this way a Gaussian in $q = k + xp$ and using the prescription (1.54) one then gets

$$\int_0^1 \int_0^\infty e^{-(\lambda(x-x^2)p^2+\lambda m^2)} \int e^{-\lambda q^2} d^D q \, \lambda \, d\lambda \, dx$$

$$= \pi^{D/2} \int_0^1 \int_0^\infty e^{-(\lambda(x-x^2)p^2+\lambda m^2)} \lambda^{-D/2} \lambda \, d\lambda \, dx$$

$$= \pi^{D/2} \Gamma(2 - D/2) \int_0^1 ((x - x^2)p^2 + m^2)^{D/2-2} \, dx,$$

which makes sense for $D \in \mathbb{C}^*$ and shows the presence of a pole at $D = 6$.

It seems then that one could simply cure these divergences by removing the polar part of the Laurent series obtained by dimensional regularization. Slightly more complicated examples with nested divergent graphs (see [Connes and Marcolli (2008)] [Collins (1986)]) show why a renormalization procedure is indeed needed at this point. In fact, the main point is that one would like to cancel the divergence by correcting the coefficients of the Lagrangian by functions of z, the *counterterms*, that is, by redoing the Feynman integral computation for the modified Lagrangian

$$\mathcal{L}(\phi) = \frac{1}{2}(\partial\phi)^2(1 - \delta Z) + \left(\frac{m^2 - \delta m^2}{2}\right)\phi^2 - \frac{g + \delta g}{6}\phi^3.$$

While one can check that this works for the example given here above, it does not seem to work any longer for more complicated examples where the graph Γ contains divergent subgraphs. In such cases, one needs to account for the way the divergences of the subgraphs have already been renormalized, by correcting the Laurent series $U(\Gamma)$ by a linear combination of other Laurent series associated to the subgraphs. This will be explained in more detail in §5.1 below.

We conclude this introductory section by recalling more formally the main distinction we already mentioned between renormalizable and non-renormalizable theories.

Definition 1.8.1. A quantum field theory with Lagrangian $\mathcal{L}(\phi) = \frac{1}{2}(\partial\phi)^2 + \frac{m^2}{2}\phi^2 + \mathcal{P}(\phi)$, with polynomial interaction terms $\mathcal{P}(\phi)$ defines a renormalizable quantum field theory if all the divergences arising from the corresponding Feynman integrals $U(\Gamma)$ can be corrected by repeatedly altering the bare constants of the existing terms in the Lagrangian or by adding a finite number of new polynomial terms to $\mathcal{L}(\phi)$.

In the case of a non-renormalizable theory, one can still follow the same renormalization procedure. However, since infinitely many new terms will have to be added to the Lagrangian to correct for the divergences of increasingly complicated graphs, the theory will end up depending on infinitely many parameters.

Chapter 2

Motives and periods

After reviewing the main tools from quantum field theory that we need, we give here a very brief and sketchy account of the other subject whose relation to quantum field theory will then be the main topic of the rest of the book, the theory of motives of algebraic varieties. This will be a simple introduction, were we will inevitably avoid the deeper and more difficult aspects of the theory, and focus only on the minimal amount of information that is necessary in order to explain how the connection to quantum field theory arises. We first sketch the construction and main properties of pure motives. As will become clearer in the following chapter, these do not suffice for the quantum field theoretic applications, for which the more complicated theory of mixed motives is needed. We are even sketchier when it comes to describing the world of mixed motives, but we give some concrete examples of mixed Tate motives, which will be the type of motives that, at least conjecturally, mainly arise in quantum field theory. We describe briefly the role of periods, which will be the main point of contact with Feynman integral computations, with special attention to the case of multiple zeta values. We also show how one can estimate the complexity of a motive through an associated invariant, a universal Euler characteristic given by the class in the Grothendieck ring of varieties. Finally, we recall the role of Galois symmetries in categories of motives.

2.1 The idea of motives

Grothendieck introduced the idea of *motives* as a universal cohomology theory for algebraic varieties. There are different notions of cohomology that can be defined for algebraic varieties: Betti cohomology, algebraic de Rham cohomology, étale cohomology. There are relations between them, like the

25

period isomorphism between de Rham and Betti cohomologies. The idea of a universal cohomology that lies behind all of these different realizations asserts that properties which occur in all the reasonable cohomologies and relations between them should come from properties that are motivic, *i.e.* that already exist at the level of this universal cohomology. Setting up a theory of motives is no easy task. The field is in fact still a very active and fast developing area of algebraic and arithmetic geometry. For a long time the only available written reference giving a detailed account of the theory of motives, along with some interesting algebro-geometric applications, was [Manin (1968)], which is still an excellent short introduction to the subject. A more recent reference that gives a broad survey of the current state of the field is [André (2004)]. An extensive series of papers on motives is available in the proceedings of the 1991 Seattle conference, collected in the two volumes [Jannsen, Kleiman, Serre (1994)]. For our purposes we will try to follow the shortest path through this intricate theory that will allow us to draw the connection to quantum field theory and discuss the main problems currently being investigated. Thus, this will be by no account an accurate introduction to motives and we refer the readers interested in a more serious and detailed account of the current state of research in this area to the recent book [André (2004)].

The theory of motives comes in two flavors, roughly corresponding to the distinction between working with smooth projective varieties or with everything else, meaning varieties that may be singular, or "non-compact". In the first case, where one only considers motives of smooth projective varieties, one obtains the theory of *pure motives*, which is built by first abstracting the main properties that a "good" cohomology theory of such algebraic varieties should satisfy (Weil cohomologies) and then constructing a category whose objects are smooth projective varieties together with the data of a projector and a Tate twist, and where the morphisms are no longer just given by morphisms of varieties but by more general *correspondences* given by algebraic cycles in the product. Under a suitable choice of an equivalence relation on the set of algebraic cycles, one can obtain in this way an abelian category of pure motives and in fact a Tannakian category, with a corresponding Galois group of symmetries, the motivic Galois group. This part of the theory of motives is at this point on solid ground, although problems like explicitly identifying the motivic Galois groups of various significant subcategories of the category of pure motives, or identifying when a correspondence is realized by an algebraic cycle, remain extremely challenging. While the general theory of pure motives is a beautiful and well established part of arithmetic algebraic geometry, the main problem of establishing Grothendieck's *standard conjectures* remains at this time open.

The situation is far more involved when it comes to *mixed motives*, namely motives associated to varieties that are no longer necessarily smooth and projective. The first issue in this case is that of identifying the right set of properties that a good cohomology theory is expected to have and that should then be also associated to a good category of motives. As it happens in topology, also in the context of algebraic varieties passing to the "non-compact" case profoundly alters the main properties of the cohomology, with Poincaré duality being replaced by the pairing of cohomologies and cohomologies with compact support, and with a prominent role played by Mayer–Vietoris type long exact sequences. The theory of mixed motives, in the form which is presently available after the work of Voevodsky, is based on triangulated categories, where these types of cohomology exact sequences manifest themselves in the form of distinguished triangles. Only in very special cases, and for very particular subcategories of the triangulated category of mixed motives, is it possible to extract from the triangulated structure a *heart* giving rise to an abelian category. This is the case for the subcategory of mixed motives that is most directly of interest to quantum field theory, namely the category of *mixed Tate motives* over a number field. As we are going to see in the next chapter, the reason why it is mixed motives instead of the simpler pure motives that arise in the quantum field theoretic context lies in the fact that certain algebraic varieties naturally associated to Feynman graphs are typically singular, so that they cannot be described, at the motivic level, using only pure motives.

When one is studying the motivic nature of certain classes of varieties, as will be the case here with certain hypersurfaces associated to Feynman graphs, it is often too compicated to work directly in the triangulated category of mixed motives, but one can sometime get useful information about the motives by looking at their classes in the *Grothendieck ring*. This consists of isomorphism classes of varieties up to equivalence relations induced by very rough "cut-and-paste" operations on algebraic varieties, with the same inclusion-exclusion property of an Euler characteristic.

2.2 Pure motives

The main properties of a good cohomology theory of algebraic varieties are summarized by the concept of Weil cohomology. We assume we are working with smooth projective varieties over a field $\mathbb{K}$, which for our purposes we can assume is $\mathbb{Q}$ or a finite algebraic extension (a number field). This means that, at each archimedean place of the number field, such a variety X defines a smooth complex manifold $X(\mathbb{C})$, which is embedded in some sufficiently large projective space $\mathbb{P}^N$. The axioms of a Weil cohomology are the following.

- $H^{\cdot}$ is a contravariant functor from the category $\mathcal{V}_{\mathbb{K}}$ of smooth projective varieties to the category $\mathrm{GrVect}_{\mathbb{K}}$ of graded finite dimensional $\mathbb{K}$-vector spaces.
- $H^i(X) = 0$ outside of the range $i \in \{0, \dots, 2\dim(X)\}$.
- Poincaré duality holds: $H^{2\dim(X)}(X) = \mathbb{K}$ and there is a non-degenerate bilinear pairing

$$\langle \cdot, \cdot \rangle : H^i(X) \times H^{2\dim(X)-i}(X) \xrightarrow{\sim} \mathbb{K}. \qquad (2.1)$$

- The Künneth formula holds:

$$H^n(X_1 \times_{\mathbb{K}} X_2) = \bigoplus_{i+j=n} H^i(X_1) \otimes_{\mathbb{K}} H_j(X_2), \qquad (2.2)$$

with the isomorphism induced by the projection maps $\pi_i : X_1 \times_{\mathbb{K}} X_2 \to X_i$, for $i = 1, 2$.

- Let $\mathfrak{Z}^i(X)$ denote the abelian group of algebraic cycles in X (smooth connected subvarieties) of codimension i. Then there exists a *cycle map*

$$\gamma_X^i : \mathfrak{Z}^i(X) \to H^{2i}(X), \qquad (2.3)$$

which is functorial with respect to pullbacks and pushforwards and satisfying

$$\gamma_{X \times Y}^n(Z \times W) = \gamma_X^i(Z) \otimes \gamma_Y^j(W),$$

with $i + j = n$, and normalized by $\gamma_{pt}^i : \mathbb{Z} \hookrightarrow \mathbb{K}$ being the standard inclusion.

- The weak Lefschetz isomorphisms hold: if H denotes a smooth hyperplane section of X, with $\iota : H \hookrightarrow X$ the inclusion, then $\iota^* : H^i(X) \to H^i(H)$ is an isomorphism for $i \le \dim(X) - 2$ and injective for $i = \dim(X) - 1$.

- The hard Lefschetz isomorphisms hold: let $\mathcal{L} : H^i(X) \to H^{i+2}(X)$ be given by $\mathcal{L} : x \mapsto x \cup \gamma_X^1(H)$. Then $\mathcal{L}^{\dim(X)-i} : H^i(X) \to H^{2\dim(X)-i}(X)$ is an isomorphism for all $i \leq \dim(X)$.

Algebraic de Rham cohomology $H^*_{dR}(X, \mathbb{K})$ and Betti (singular) cohomology $H^*_B(X) = H^*_{sing}(X(\mathbb{C}), \mathbb{Q})$ are both examples of Weil cohomologies. It is in general extremely difficult to determine whether a given class in a Weil cohomology $H^*(X)$ lies in the image of the cycle map γ_X^*. The Hodge conjecture, for example, is closely related to this question. The fact that in general algebraic cycles are very complicated objects and one does not know much about how to characterize them homologically constitutes the main difficulty in the theory of pure motives, as the construction of the category of pure motives that we now recall will show more clearly.

Grothendieck's idea that gave birth to the whole theory of motives was that there should be a cohomology theory for (smooth projective) algebraic varieties that should map via *realization* functors to any cohomology satisfying the axioms recalled above. The existence of isomorphisms between the different known Weil cohomologies, such as the *period isomorphism* between Betti and de Rham cohomology, motivated this search for an underlying universal theory.

The approach to constructing such a universal cohomology theory is to remain sufficiently close to the geometry of the varieties themselves, instead of passing to associated vector spaces or modules over a ring, as cohomologies typically do. One modifies instead the maps between varieties so as to obtain a category that still contains the usual category of smooth projective varieties. It has additional objects and morphisms that make it into a "linearization" of the category of varieties, that is, a category that is good enough to behave as a category of modules over a ring.

The precise notion for a category that behaves like a category of modules over a ring, namely a category that is good enough to do homological algebra, is that of an *abelian category*, see [Gelfand and Manin (1994)].

Definition 2.2.1. An abelian category $\mathcal{C}$ is a category satisfying the following properties.

- For all $X, Y \in \mathrm{Obj}(\mathcal{C})$ the set of morphisms $\mathrm{Hom}_{\mathcal{C}}(X, Y)$ is an abelian group (*i.e.* $\mathcal{C}$ is pre-additive). The composition of morphisms is compatible with the abelian group structure of the Hom-sets.
- There is a zero-object $0 \in \mathrm{Obj}(\mathcal{C})$ with $\mathrm{Hom}_{\mathcal{C}}(0,0) = 0$, the trivial abelian group.

- There are finite products and coproducts in $\mathcal{C}$.
- All morphisms have kernels and cokernels in the category and every morphism $f : X \to Y$ has a canonical decomposition

$$K \xrightarrow{k} X \xrightarrow{i} I \xrightarrow{j} Y \xrightarrow{c} K', \qquad (2.4)$$

with $j \circ i = f$, and with $K = \mathrm{Ker}(f) \in \mathrm{Obj}(\mathcal{C})$, $K' = \mathrm{Coker}(f) \in \mathrm{Obj}(\mathcal{C})$, and $I = \mathrm{Coker}(k) = \mathrm{Ker}(c) \in \mathrm{Obj}(\mathcal{C})$.

The property that an abelian category "behaves like a category of modules over a ring" is made precise by the Freyd–Mitchell theorem, which shows that indeed a *small* abelian category can be faithfully embedded as a subcategory of a category of modules over a ring.

To obtain a category of pure motives one then aims at constructing an abelian category of smooth projective varieties and correspondences, which admits realization functors to vector spaces determined by the various possible Weil cohomologies.

A morphism $f : X \to Y$ of varieties can be regarded as a correspondence by considering its graph

$$\mathcal{G}(f) = \{(x, y) \in X \times Y \,|\, y = f(x)\} \subset X \times Y. \qquad (2.5)$$

This is a closed connected subvariety of the product $X \times Y$ of codimension

$$\mathrm{codim}(\mathcal{G}(f)) = \dim(Y). \qquad (2.6)$$

Since one is aiming at constructing a universal *co*-homology theory, which should be a contravariant functor, we regard the graph $\mathcal{G}(f)$ of a morphism $f : X \to Y$ as a correspondence from Y to X,

$$\mathcal{G}(f) \in \mathrm{Corr}^{\dim(Y)}(Y, X), \qquad (2.7)$$

where $\mathrm{Corr}^i(Y, X)$ is the abelian group of correspondences from Y to X given by algebraic cycles in $\mathfrak{Z}^i(X \times Y)$. Thus, the generalization of morphisms $f : X \to Y$ is given by correspondences of the form $Z = \sum_j n_j Z_j$, that are finite linear combinations with $n_j \in \mathbb{Z}$ of irreducible smooth algebraic subvarieties $Z_j \subset X \times Y$. The usual composition of morphisms

$$X_1 \xrightarrow{f_1} X_2 \xrightarrow{f_2} X_3$$

is replaced by a composition of correspondences given by the *fiber product*

$$Z' \circ Z = Z' \times_{X_2} Z, \qquad (2.8)$$

for $Z \subset X_1 \times X_2$ and $Z' \subset X_2 \times X_3$. This means that, to obtain the composition $Z' \circ Z$ one does the following operations:

- Take the preimages $\pi_{32}^{-1}(Z')$ and $\pi_{21}^{-1}(Z)$ in $X_1 \times X_2 \times X_3$, with π_{ij} : $X_1 \times X_2 \times X_3 \to X_i \times X_j$ the projections.
- Intersect $\pi_{32}^{-1}(Z') \cap \pi_{21}^{-1}(Z)$ in $X_1 \times X_2 \times X_3$.
- Push forward the result along the remaining projection $\pi_{31} : X_1 \times X_2 \times X_3 \to X_1 \times X_3$,

$$Z' \circ Z = (\pi_{31})_*(\pi_{32}^{-1}(Z') \cap \pi_{21}^{-1}(Z)) \subset X_1 \times X_3. \qquad (2.9)$$

- Extend by linearity to correspondences of the form $Z = \sum_j n_j Z_j$ and $Z' = \sum_r n_r Z_r'$.

Because we are working with algebraic varieties as opposed to smooth manifolds, one in general cannot simply deform them to be in general position so that the intersection $\pi_{32}^{-1}(Z') \cap \pi_{21}^{-1}(Z)$ is transverse. Thus, one needs to use some more sophisticated tools like excess intersection to deal with intersection products in non-transverse situations, as one does in algebraic geometry to define the product in the Chow ring, and use an equivalence relation on the set of algebraic cycles that preserves the intersection product and allows for different choices of representatives within the same class. Reducing the size of the groups of algebraic cycles by an equivalence relation is also desirable to obtain groups that are, at least in some cases, finite dimensional.

There is more than one possible choice of an equivalence relation on algebraic cycles, and the properties of the resulting category of motives change according to which equivalence relation one considers. We only mention here *numerical* and *homological* equivalence. We refer the reader to [André (2004)] for a discussion of *rational* equivalence and the corresponding properties. Imposing the numerical equivalence relation means identifying two cycles $Z_1 \sim_{num} Z_2$ if their intersection products agree, $\#(Z_1 \cap Z) = \#(Z_2 \cap Z)$, for arbitrary subvarieties Z of complementary dimension. The homological equivalence requires the choice of a Weil cohomology H^* and prescribes that a cycle $Z \in \mathfrak{Z}^i(X \times Y)$ is homologically trivial, $Z \sim_{hom} 0$, if its image under the cycle map $\gamma_{X \times Y}^i : \mathfrak{Z}^i(X \times Y) \to H^{2i}(X \times Y)$ is trivial, $\gamma_{X \times Y}^i(Z) = 0$. Although this second relation depends on the choice of a Weil cohomology, it is part of Grothendieck's standard conjectures that numerical and homological equivalence agree, thus making the latter independent of the choice of the cohomology. One denotes by $\mathrm{Corr}_\sim^i(X \times Y)$ the abelian group generated by the algebraic cycles of codimension i in $X \times Y$ modulo the equivalence relation $\sim$. This is compatible with the composition of correspondences defined using the fiber product. We also use the notation

$$\mathfrak{C}^0_\sim(X,Y) := \mathrm{Corr}^{\dim(X)}_\sim(Y \times X) \otimes_{\mathbb{Z}} \mathbb{Q}. \qquad (2.10)$$

One can then define a category of motives as follows.

Definition 2.2.2. The category $\mathcal{M}^{\mathrm{eff}}_{num}(\mathbb{K})$ of effective numerical pure motives has as objects pairs (X,p), where X is a smooth projective variety over $\mathbb{K}$ and $p \in \mathfrak{C}^0_{num}(X,X)$ is an idempotent, $p^2 = p$. The morphisms are given by

$$\mathrm{Hom}_{\mathcal{M}^{\mathrm{eff}}_{num}(\mathbb{K})}((X,p),(Y,q)) = q\mathfrak{C}^0_\sim(X,Y)p. \qquad (2.11)$$

The reason for introducing projectors $p^2 = p$ is in order to be able to "cut out" pieces of the cohomology of a variety X, for example the piece that corresponds to a certain degree only. It is in general very difficult to give an explicit description in terms of algebraic cycles of a projector that isolates a given component of the cohomology of a variety. Notice also that, since $p \in \mathfrak{C}^0_\sim(X,X)$, the property of being idempotent also depends on the choice of the equivalence relation. The set of morphisms here are $\mathbb{Q}$-vector spaces instead of abelian groups (the category is $\mathbb{Q}$-linear).

An important recent result [Jannsen (1992)] shows that the category $\mathcal{M}^{\mathrm{eff}}_{num}(\mathbb{K})$ is a *semisimple abelian category*. The semisimplicity property means that there exists a set $\mathcal{S}$ of objects such that for all $X \in \mathcal{S}$ one has $\mathrm{Hom}(X,X) = \mathbb{Q}$ and $\mathrm{Hom}(X,Y) = 0$ for all $X \neq Y \in \mathcal{S}$ (simple objects), and such that every object X of the category has a decomposition $X = \oplus_i X_i$ with summands $X_i \in \mathcal{S}$.

The category $\mathcal{M}^{\mathrm{eff}}_{num}(\mathbb{K})$ has a tensor product that comes form the product of varieties. This makes it into a tensor category. A desirable property for a tensor category is a duality that makes it into a *rigid* tensor category.

Recall first that a category $\mathcal{C}$ where the set $\mathrm{Hom}_{\mathcal{C}}(X,Y)$ is a $\mathbb{K}$-vector space, for any $X,Y \in \mathrm{Obj}(\mathcal{C})$, is called $\mathbb{K}$-linear. A tensor category over $\mathbb{K}$ is a $\mathbb{K}$-linear category $\mathcal{C}$, with a bi-functor $\otimes : \mathcal{C} \times \mathcal{C} \to \mathcal{C}$ and an object $1 \in \mathrm{Obj}(\mathcal{C})$ for which there are functorial isomorphisms

$$a_{X,Y,Z} : X \otimes (Y \otimes Z) \to (X \otimes Y) \otimes Z$$

$$c_{X,Y} : X \otimes Y \to Y \otimes X$$

$$l_X : X \otimes 1 \to X \quad \text{and} \quad r_X : 1 \otimes X \to X.$$

These satisfy consistency conditions expressed in the form of a triangle, a pentagon, and a hexagon diagram, see [André (2004)] for details. One also generally assumes that commutativity holds, which means that $c_{Y,X} = c_{X,Y}^{-1}$, although this can be replaced by a signed version.

Definition 2.2.3. A $\mathbb{K}$-linear tensor category $\mathcal{C}$ is rigid if there is a duality $\vee : \mathcal{C} \to \mathcal{C}^{op}$ such that, for all $X \in \mathrm{Obj}(\mathcal{C})$ the functor $\cdot \otimes X^\vee$ is left adjoint to $\cdot \otimes X$ and $X^\vee \otimes \cdot$ is right adjoint to $X \otimes \cdot$. Moreover, there are an evaluation $\epsilon : X \otimes X^\vee \to 1$ (the unit object for tensor product) and a unit $\delta : 1 \to X^\vee \otimes X$ satisfying

$$(\epsilon \otimes 1) \circ (1 \otimes \delta) = 1_X, \quad (1 \otimes \epsilon) \circ (\delta \otimes 1) = 1_{X^\vee}.$$

One also assumes that $\mathrm{Hom}_\mathcal{C}(1,1) = \mathbb{K}$.

Recall here that given two functors $F : \mathcal{C} \to \mathcal{C}'$ and $G : \mathcal{C}' \to \mathcal{C}$ satisfying

$$\mathrm{Hom}_\mathcal{C}(F(X),Y) = \mathrm{Hom}_{\mathcal{C}'}(X,G(Y)),$$

for all $X \in \mathrm{Obj}(\mathcal{C})$ and $Y \in \mathrm{Obj}(\mathcal{C}')$, with the identification realized by a natural isomorphism, one says that F is left-adjoint to G and G is right-adjoint to F.

The category of numerical effective pure motives does not have a duality. One can see that, to have such a duality, one would need to take the transpose of a correspondence $Z \subset X \times Y$, but this would have the effect of changing the codimension. Namely, if $Z \in \mathfrak{C}^0_{num}(X,Y) = \mathrm{Corr}_{num}^{\dim(X)}(Y \times X) \otimes_{\mathbb{Z}} \mathbb{Q}$, then the transpose $Z^\dagger$ is in $\mathrm{Corr}_{num}^{\dim(X)}(X \times Y) \otimes_{\mathbb{Z}} \mathbb{Q}$, but the latter is not $\mathfrak{C}^0_{num}(Y,X) = \mathrm{Corr}_{num}^{\dim(Y)}(X \times Y) \otimes_{\mathbb{Z}} \mathbb{Q}$. Thus, to have a duality one needs to include among morphisms also cycles of other codimensions. This can be achieved by introducing new objects, the *Tate motives* $\mathbb{Q}(n)$. One defines the Tate motive $\mathbb{Q}(1)$ to be the formal inverse of the Lefschetz motive $\mathbb{L}$. This is the motive whose cohomological realization is $H^2(\mathbb{P}^1)$. This means that one writes the motive of $\mathbb{P}^1$ as $1 \oplus \mathbb{L}$, where 1 is the motive of a point, which corresponds to the cohomology $H^0(\mathbb{P}^1)$. Similarly, one writes the motive of $\mathbb{P}^n$ as $1 \oplus \mathbb{L} \oplus \mathbb{L}^2 \oplus \cdots \oplus \mathbb{L}^n$, which corresponds to the decomposition of $\mathbb{P}^n$ into cells $\mathbb{A}^0 \cup \mathbb{A}^1 \cup \cdots \cup \mathbb{A}^n$, and the corresponding cohomology ring $H^*(\mathbb{P}^n; \mathbb{Z}) = \mathbb{Z}[u]/u^{n+1}$ with the generator u in degree 2. As an object in the category $\mathcal{M}_{num}(\mathbb{K})$ (see Definition 2.2.4 below), the Lefschetz motive is given by the triple $\mathbb{L} = (\mathrm{Spec}(\mathbb{K}), id, -1)$, with $1 = (\mathrm{Spec}(\mathbb{K}), id, 0)$ being the motive of a point. Introducing the formal inverse $\mathbb{Q}(1) = \mathbb{L}^{-1} = (\mathrm{Spec}(\mathbb{K}), id, 1)$, with $\mathbb{Q}(n) = \mathbb{Q}(1)^{\otimes n}$ and $\mathbb{Q}(0) = 1$, has the effect of shifting the codimensions of cycles in the definitions of morphisms in the category of motives in the following way.

Definition 2.2.4. The category $\mathcal{M}_{num}(\mathbb{K})$ has objects given by triples (X, p, m) with $(X, p) \in \mathrm{Obj}(\mathcal{M}_{num}^{\mathrm{eff}}(\mathbb{K}))$ an effective numerical motive and $m \in \mathbb{Z}$. The morphisms are given by

$$\mathrm{Hom}_{\mathcal{M}_{num}(\mathbb{K})}((X, p, m), (Y, q, n)) = q\mathfrak{C}^{n-m}(X, Y)p, \qquad (2.12)$$

where

$$\mathfrak{C}^{m-n}(X, Y) = \mathrm{Corr}_{num}^{\dim(X)+n-m}(Y \times X) \otimes_{\mathbb{Z}} \mathbb{Q} \qquad (2.13)$$

The tensor product is then given by

$$(X, p, m) \otimes (Y, q, n) = (X \times Y, p \otimes q, n + m), \qquad (2.14)$$

and the duality is of the form

$$(X, p, m)^{\vee} = (X, p^{\dagger}, \dim(X) - m). \qquad (2.15)$$

One obtains in this way a rigid tensor category. As we discuss in §2.9 below, the resulting category is in fact more than a rigid tensor category: it has enough structure to be identified uniquely as the category of finite dimensional linear representations of a group (more precisely of an affine group scheme) via the *Tannakian formalism*, see §2.9 below.

2.3 Mixed motives and triangulated categories

Having just quickly mentioned how one constructs a good category of pure motives, we jump ahead and leave the world of smooth projective varieties to venture into the more complicated and more mysterious realm of mixed motives. One should think of mixed motives as objects that come endowed with filtrations whose graded pieces are pure motives. The difficult nature of mixed motives therefore lies in the nontrivial extensions of pure motives. From the cohomological point of view, instead of working with Weil cohomologies, one considers a different set of properties that a universal cohomology theory should satisfy in this case, where the varieties involved may be non-compact or singular. The main properties that one expects such a cohomology theory to satisfy become of the following form:

- Homotopy invariance: the projection $\pi : X \times \mathbb{A}^1 \to X$ induces an isomorphism

$$\pi^* : H^*(X) \xrightarrow{\simeq} H^*(X \times \mathbb{A}^1). \qquad (2.16)$$

- Mayer–Vietoris sequence: if U, V are an open covering of X then there is a long exact cohomology sequence

$$\cdots \to H^{i-1}(U \cap V) \to H^i(X) \to H^i(U) \oplus H^i(V) \to H^i(U \cap V) \to \cdots$$
$$(2.17)$$

- Poincaré duality is replaced by the duality between cohomology H^* and cohomology with compact support, H_c^*.
- The Künneth formula still holds.

This replaces the notion of a Weil cohomology with "mixed Weil cohomologies". Ideally, one would like to have an abelian category $\mathcal{MM}(\mathbb{K})$ of mixed motives that has realization functors to all the mixed Weil cohomologies. This is (at present) not available, although after a long series of developments and deep results in the field (see [Voevodsky (2000)], [Levine (1998)], [Hanamura (1995)]) one has at least a weaker version available, in the form of a *triangulated* category of mixed motives $\mathcal{DM}(\mathbb{K})$.

Triangulated categories still allow for several techniques from homological algebra, in particular in the form of long exact cohomology sequences.

Definition 2.3.1. A triangulated category $\mathcal{T}$ is an additive category endowed with an automorphism $T : \mathcal{T} \to \mathcal{T}$ (the shift functor) and a family of distinguished triangles

$$X \to Y \to Z \to T(X) \qquad (2.18)$$

with the following properties:

- $X \xrightarrow{id} X \to 0 \to T(X)$ is a distinguished triangle for all $X \in \mathrm{Obj}(\mathcal{T})$.
- For all $f \in \mathrm{Hom}_{\mathcal{T}}(X, Y)$, there exists an object $C(f) \in \mathrm{Obj}(\mathcal{T})$ (the mapping cone) with a morphism $C(f) \to T(X)$, such that $X \xrightarrow{f} Y \to C(f) \to T(X)$ is a distinguished triangle.
- A triangle isomorphic to a distinguished triangle is distinguished (where isomorphisms of triangles are isomorphisms of the objects such that the resulting diagram commutes).
- Rotations $Y \to Z \to T(X) \to T(Y)$ and $T^{-1}(Z) \to X \to Y \to Z$ of a distinguished triangle $X \to Y \to Z \to T(X)$ are distinguished.
- Given distinguished triangles $X \to Y \to Z \to T(X)$ and $X' \to Y' \to Z' \to T(X')$ and morphisms $f : X \to X'$ and $h : Y \to Y'$ such that the square commutes, there exists a morphism $Z \to Z'$ such that all squares commute, with $T(f) : T(X) \to T(X')$ completing the diagram.
- Given $X \xrightarrow{f} Y \to C(f) \to T(X)$ and $Y \xrightarrow{h} Z \to C(h) \to T(Y)$, and the composite $X \xrightarrow{h \circ f} Z \to C(h \circ f) \to T(X)$, these three triangles can

be placed on the vertices and edges of an octahedron with morphisms completing the picture so that all the faces are commutative diagrams.

One typically writes $X[1]$ for $T(X)$ and $X[n] = T^n(X)$ for $n \in \mathbb{Z}$.

One more property of triangulated categories that follows from these axioms, and which we will need explicitly later, is the following.

Lemma 2.3.2. *Given a triangulated category T, a full subcategory T' is a triangulated subcategory if and only if it is invariant under the shift T of T and for any distinguished triangle $X \to Y \to Z \to X[1]$ for T where X and Y are in T' there is an isomorphism $Z \simeq Z'$ where Z' is also an object in T'.*

A full triangulated subcategory $T' \subset T$ is *thick* if it is closed under direct sums.

It is sometimes possible to extract an abelian category from a triangulated category using the method of t-structures.

Definition 2.3.3. A t-structure on a triangulated category T consists of two full subcategories $T^{\leq 0}$ and $T^{\geq 0}$ with the following property. Upon denoting by $T^{\geq n} = T^{\geq 0}[-n]$ and $T^{\leq n} = T^{\leq 0}[-n]$, one has

$$T^{\leq -1} \subset T^{\leq 0}, \quad \text{and} \quad T^{\geq 1} \subset T^{\geq 0}$$

with the property that

$$\mathrm{Hom}_T(X, Y) = 0, \quad \forall X \in \mathrm{Obj}(T^{\leq 0}), \quad \forall Y \in \mathrm{Obj}(T^{\geq 1}),$$

and such that, for all $Y \in \mathrm{Obj}(T)$, there exists a distinguished triangle

$$X \to Y \to Z \to X[1]$$

with $X \in \mathrm{Obj}(T^{\leq 0})$ and $Z \in \mathrm{Obj}(T^{\geq 1})$.

When one has a t-structure on a triangulated category, one can consider the full subcategory $T^0 = T^{\leq 0} \cap T^{\geq 0}$. This is called the *heart of the t-structure* and it is known to be an abelian category, see [Beĭlinson, Bernstein, Deligne (1982)], [Kashiwara and Schapira (1990)].

Theoretical physicists have recently become familiar with the formalism of t-structures in triangulated categories in the context of homological mirror symmetry, see for instance [Bridgeland (2009)].

2.4 Motivic sheaves

We also mention briefly, for later use, a recent construction due to [Arapura (2008)] of a category of motivic sheaves over a base scheme S, modeled on Nori's approach to the construction of categories of mixed motives. We will return in Chapter 7 to discuss how this fits with the motives associated to Feynman diagrams.

The category of motivic sheaves constructed in [Arapura (2008)] is based on Nori's construction of categories of motives via representations of graphs made up of objects and morphisms (*cf.* [Bruguières (2004)]).

According to [Arapura (2008)], one constructs a category of motivic sheaves over a scheme S, by taking as vertices of the corresponding graph objects of the form

$$(f : X \to S, Y, i, w), \qquad (2.19)$$

where $f : X \to S$ is a quasi-projective morphism, $Y \subset X$ is a closed subvariety, $i \in \mathbb{N}$, and $w \in \mathbb{Z}$. One thinks of such an object as determining a motivic version $h_S^i(X, Y)(w)$ of the local system given by the (Tate twisted) fiberwise cohomology of the pair $H_S^i(X, Y; \mathbb{Q}) = R^i f_* j_! \mathbb{Q}_{X \setminus Y}$, where $j = j_{X \setminus Y} : X \setminus Y \hookrightarrow X$ is the open inclusion, *i.e.* the sheaf defined by

$$U \mapsto H^i(f^{-1}(U), f^{-1}(U) \cap Y; \mathbb{Q}).$$

The edges are given by the geometric morphisms:

- morphisms of varieties over S,

$$(f_1 : X_1 \to S, Y_1, i, w) \to (f_2 : X_2 \to S, Y_2 = F(Y), i, w), \qquad (2.20)$$

with $f_2 \circ F = f_1$;
- the connecting morphisms

$$(f : X \to S, Y, i+1, w) \to (f|_Y : Y \to S, Z, i, w), \quad \text{for} \quad Z \subset Y \subset X; \qquad (2.21)$$

- the twisted projection morphisms

$$(f : X \times \mathbb{P}^1 \to S, Y \times \mathbb{P}^1 \cup X \times \{0\}, i+2, w+1) \to (f : X \to S, Y, i, w). \qquad (2.22)$$

The product in this category of motivic sheaves is given by the fiber product

$$\begin{aligned} (X \to S, Y, i, w) \times (X' \to S, Y', i', w') = \\ (X \times_S X' \to S, Y \times_S X' \cup X \times_S Y', i+i', w+w'). \end{aligned} \qquad (2.23)$$

2.5   The Grothendieck ring of motives

An invariant of algebraic varieties and motives that contains a lot of information and can be used to evaluate the motivic complexity of a given algebraic variety is the class in the Grothendieck ring. This is constructed by breaking up the variety into pieces, by simple cut-and-paste operations of the scissors congruence type. The operation of assigning to a variety its class in the Grothendieck ring behaves like a universal Euler characteristic.

Definition 2.5.1. Let $\mathcal{V}_{\mathbb{K}}$ denote the category of algebraic varieties over a number field $\mathbb{K}$. The Grothendieck ring $K_0(\mathcal{V}_{\mathbb{K}})$ is the abelian group generated by isomorphism classes $[X]$ of varieties, with the relation

$$[X] = [Y] + [X \smallsetminus Y], \tag{2.24}$$

for $Y \subset X$ closed. It is made into a ring by the product $[X \times Y] = [X][Y]$.

Let $\mathcal{M}(\mathbb{K})$ denote, as above, the abelian category of (numerical) pure motives over $\mathbb{K}$. As above we write the objects of $\mathcal{M}(\mathbb{K})$ in the form (X, p, m), with X a smooth projective variety over $\mathbb{K}$, $p = p^2 \in \mathrm{End}(X)$ a projector, and $m \in \mathbb{Z}$ the Tate twist. Let $K_0(\mathcal{M}(\mathbb{K}))$ denote the Grothendieck ring of the abelian category $\mathcal{M}(\mathbb{K})$.

We'll see that the class of a variety, or of a motive, in the corresponding Grothendieck ring, can be thought of as a "universal Euler characteristic". To this purpose, we first recall the notion of *additive invariants* of varieties.

Definition 2.5.2. An *additive invariant* is a map $\chi : \mathcal{V}_{\mathbb{K}} \to R$, with values in a commutative ring R, satisfying the following properties.

- Isomorphism invariance: $\chi(X) = \chi(Y)$ if $X \cong Y$ are isomorphic.
- Inclusion-exclusion: $\chi(X) = \chi(Y) + \chi(X \smallsetminus Y)$, for a closed embedding $Y \subset X$.
- Multiplicative property: $\chi(X \times Y) = \chi(X)\chi(Y)$.

The topological Euler characteristic clearly satisfies all of these properties and is in fact the prototypical example of such an invariant. Counting points over finite fields is another invariant that behaves in the same way and factors through the Grothendieck ring.

By the definition of the relations on the Grothendieck ring one sees that assigning an additive invariant with values in R is equivalent to assigning a ring homomorphism $\chi : K_0(\mathcal{V}_K) \to R$.

The results of [Gillet and Soulé (1996)] show that there exists an additive invariant $\chi_{mot} : \mathcal{V}_{\mathbb{K}} \to K_0(\mathcal{M}(\mathbb{K}))$. This assigns to a smooth projective variety X the class $\chi_{mot}(X) = [(X, id, 0)] \in K_0(\mathcal{M}(\mathbb{K}))$. The value assigned to a more general variety is more complicated to describe and it is given in [Gillet and Soulé (1996)] in terms of an object $W(X)$ in the category of complexes over $\mathcal{M}(\mathbb{K})$. (Technically, here the category $\mathcal{M}(\mathbb{K})$ is taken with the rational equivalence relation on algebraic cycles.) It is sometime referred to as "motivic Euler characteristic".

If $\mathbb{L}$ denotes the class $\mathbb{L} = [\mathbb{A}^1] \in K_0(\mathcal{V}_{\mathbb{K}})$ then its image in $K_0(\mathcal{M}(\mathbb{K}))$ is the Lefschetz motive $\mathbb{L} = \mathbb{Q}(-1) = [(\mathrm{Spec}(K), id, -1)]$. Since the Lefschetz motive is invertible in $K_0(\mathcal{M}(\mathbb{K}))$, its inverse being the Tate motive $\mathbb{Q}(1)$, the ring homomorphism $\chi_{mot} : \mathcal{V}_{\mathbb{K}} \to K_0(\mathcal{M}(\mathbb{K}))$ induces a ring homomorphism

$$\chi_{mot} : K_0(\mathcal{V}_{\mathbb{K}})[\mathbb{L}^{-1}] \to K_0(\mathcal{M}(\mathbb{K})). \qquad (2.25)$$

This in particular means that one can work either in the Grothendieck ring of varieties or of motives, being able to map the first to the latter through χ_{mot}. In the following, we will usually refer to the Grothendieck ring $K_0(\mathcal{V}_{\mathbb{K}})$. A particularly interesting subring is the one generated by the Lefschetz motive $\mathbb{L} = [\mathbb{A}^1]$. This is a polynomial ring $\mathbb{Z}[\mathbb{L}] \subset K_0(\mathcal{V}_{\mathbb{K}})$, or a ring of Laurent polynomials $\mathbb{Z}[\mathbb{L}, \mathbb{L}^{-1}] \subset K_0(\mathcal{M}(\mathbb{K}))$, which corresponds to the classes of varieties that motivically are mixed Tate motives. See the comment at the end of §2.6 below on the relation between the information given by knowing that the class in the Grothendieck ring lies in the subring $\mathbb{Z}[\mathbb{L}] \subset K_0(\mathcal{V}_{\mathbb{K}})$ and the question of whether the motive, as an object of the triangulated category $\mathcal{DM}(\mathbb{K})$ is a mixed Tate motive.

2.6 Tate motives

Let $\mathcal{DM}(\mathbb{K})$ be the Voevodsky triangulated category of mixed motives over a field $\mathbb{K}$ [Voevodsky (2000)]. The triangulated category $\mathcal{DMT}(\mathbb{K})$ of mixed Tate motives is the full triangulated thick subcategory of $\mathcal{DM}(\mathbb{K})$ generated by the Tate objects $\mathbb{Q}(n)$. It is known that, over a number field $\mathbb{K}$, there is a canonical t-structure on $\mathcal{DMT}(\mathbb{K})$ and one can therefore construct an abelian category $\mathcal{MT}(\mathbb{K})$ of mixed Tate motives (see [Levine (1998)]).

For the kind of applications we are going to see in the rest of this book, we will be especially interested in the question of when a given mixed motive is mixed Tate. In sufficiently simple cases, one can try to answer this question using the properties of $\mathcal{DMT}(\mathbb{K})$ as a triangulated subcategory of the triangulated category $\mathcal{DM}(\mathbb{K})$ of mixed motives. Although the available explicit constructions of $\mathcal{DM}(\mathbb{K})$ given in [Voevodsky (2000)], [Levine (1998)], [Hanamura (1995)], are technically very complicated and certainly beyond the scope of this book, often when dealing with very specific concrete examples, one does not need the full strength of the general construction, but one can directly use the formal properties that the triangulated category $\mathcal{DM}(\mathbb{K})$ satisfies. In particular, in order to understand whether the motive of a given variety is mixed Tate, one can try to find a geometric decomposition, often in the form of a stratification, of the variety, in such a way that one has enough control on the strata to build the motive out of pieces that are mixed Tate. More precisely, in such an approach one wants to make repeated use of two properties of the triangulated category $\mathcal{DM}(\mathbb{K})$:

- (Proposition 4.1.4 of [Voevodsky (2000)]) Over a field of characteristic zero, a closed embedding $Y \subset X$ determines a distinguished triangle in $\mathcal{DM}(\mathbb{K})$

$$\mathfrak{m}(Y) \to \mathfrak{m}(X) \to \mathfrak{m}(X \smallsetminus Y) \to \mathfrak{m}(Y)[1], \qquad (2.26)$$

  where the notation $\mathfrak{m}(X)$ here stands for the motive with compact support, denoted by $\underline{C}_*^c(X)$ in [Voevodsky (2000)].
- (Corollary 4.1.8 of [Voevodsky (2000)]) Homotopy invariance implies the identity of motives

$$\mathfrak{m}(X \times \mathbb{A}^1) = \mathfrak{m}(X)(1)[2], \qquad (2.27)$$

  where $[n]$ is the shift, and (n) is the Tate twist by $\mathbb{Q}(n)$.

The distinguished triangle (2.26) implies that, since $\mathcal{DMT}(\mathbb{K})$ is a triangulated subcategory of $\mathcal{DM}(\mathbb{K})$, if two of the terms are known to be in $\mathcal{DMT}(\mathbb{K})$ then the third one also is. This can often be used as a way to reduce a motive to simpler building blocks for which it may be easier to answer the question of whether they are mixed Tate or not, and then derive information about the original motive via the distinguished triangles and homotopy invariance property.

We reported above the distinguished triangles and homotopy invariance properties for the motives with compact support in Voevodsky's terminology. One can obtain analogous statements for motives without compact support using the duality in $\mathcal{DM}(\mathbb{K})$ between mixed motives and mixed motives with compact support, which replaces Poincaré duality that holds for Weil cohomologies in the pure motives setting.

We give here an example from [Aluffi and Marcolli (2009a)], which will be relevant to Feynman integrals (see §3.13 below), namely the case of *determinant hypersurfaces*. It is a well known fact that, motivically, these varieties are mixed Tate (see [Belkale and Brosnan (2003a)]), but we illustrate here explicitly how one can argue that this is the case using only the two properties listed above of the triangulated category of mixed Tate motives.

Definition 2.6.1. Let $\hat{\mathcal{D}}_\ell \subset \mathbb{A}^{\ell^2}$ be the hypersurface defined by the vanishing of the determinant

$$\hat{\mathcal{D}}_\ell = \{x = (x_{ij}) \in \mathbb{A}^{\ell^2} \mid \det(x) = 0\}. \tag{2.28}$$

Let $\mathcal{D}_\ell \subset \mathbb{P}^{\ell^2-1}$ be the corresponding projective hypersurface defined by the homogeneous polynomial $\det(x)$.

We have the following result on the motivic nature of the determinant hypersurfaces. We formulate it in terms of hypersurface complements, as that will be the natural geometric object to consider in the case of the Feynman integrals. Clearly, if the result holds for the hypersurface complement it also holds for the hypersurface itself, by the first property above on the distinguished triangle of a closed embedding, and the fact that affine and projective spaces are Tate motives.

Theorem 2.6.2. *The determinant hypersurface complement* $\mathbb{A}^{\ell^2} \smallsetminus \hat{\mathcal{D}}_\ell$ *defines an object in the category* $\mathcal{DMT}$ *of mixed Tate motives.*

Proof. Since $\hat{\mathcal{D}}_\ell$ consists of $\ell \times \ell$ matrices of rank $< \ell$, in order to describe the locus $\mathbb{A}^{\ell^2} \smallsetminus \hat{\mathcal{D}}_\ell$ we need to parameterize matrices of rank exactly equal to ℓ, which can be done in the following way. Fix an ℓ-dimensional vector space E.

- Denote by $\mathcal{W}_1$ the variety $E \smallsetminus \{0\}$;
- Note that $\mathcal{W}_1$ is equipped with a trivial vector bundle $E_1 = E \times \mathcal{W}_1$, and with a line bundle $S_1 := L_1 \subseteq E_1$ whose fiber over $v_1 \in \mathcal{W}_1$ consists of the line spanned by v_1;
- Let $\mathcal{W}_2 \subseteq E_1$ be the complement $E_1 \smallsetminus L_1$;
- Note that $\mathcal{W}_2$ is equipped with a trivial vector bundle $E_2 = E \times \mathcal{W}_2$, and *two* line subbundles of E_2: the pull-back of L_1 (still denoted L_1) and the line-bundle L_2 whose fiber over $v_2 \in \mathcal{W}_2$ consists of the line spanned by v_2;
- By construction, L_1 and L_2 span a rank-2 subbundle S_2 of E_2;

– Let $\mathcal{W}_3 \subseteq E_2$ be the complement $E_2 \smallsetminus S_2$; and so on.

We first show that the variety $\mathcal{W}_\ell$ constructed as above is isomorphic to the locus we want, namely $\mathbb{A}^{\ell^2} \smallsetminus \hat{D}_\ell$.

To see this, first observe that at the k-th step, the procedure described above produces a variety $\mathcal{W}_k$, endowed with k line bundles $L_1, \ldots, L_k$ spanning a rank-k subbundle S_k of the trivial vector bundle $E_k := E \times \mathcal{W}_k$. If $S_k \subsetneq E_k$, define $\mathcal{W}_{k+1} := E_k \smallsetminus S_k$. Let $E_{k+1} = E \times \mathcal{W}_{k+1}$, and define the line subbundles $L_1, \ldots, L_k$ to be the pull-backs of the similarly named line bundles on $\mathcal{W}_k$. Let L_{k+1} be the line bundle whose fiber over v_{k+1} is the line spanned by v_{k+1}. The line bundles $L_1, \ldots, L_{k+1}$ span a rank $k + 1$ subbundle S_{k+1} of E_{k+1}, and the construction can continue. The sequence stops at the ℓ-th step, where S_ℓ has rank ℓ, equal to the rank of E_ℓ, so that $E_\ell \smallsetminus S_\ell = \emptyset$.

Each variety $\mathcal{W}_k$ maps to $\mathbb{A}^{\ell^2}$ as follows: a point of $\mathcal{W}_k$ determines k vectors $v_1, \ldots, v_k$, and can be mapped to the matrix whose first k rows are $v_1, \ldots, v_k$ resp. (and the remaining rows are 0). By construction, this matrix has rank exactly k. Conversely, any such rank k matrix is the image of a point of $\mathcal{W}_k$, by construction.

Also notice that, in the construction described above, the bundle S_k over $\mathcal{W}_k$ is in fact trivial. An explicit trivialization is given by the map $\mathbb{K}^k \times \mathcal{W}_k \overset{\alpha}{\to} S_k$ defined by $\alpha : ((c_1, \ldots, c_r), (v_1, \ldots, v_r)) \mapsto c_1 v_1 + \cdots + c_r v_r$, where points of $\mathcal{W}_k$ are parameterized by k-tuples of vectors $v_1, \ldots, v_k$ spanning $S_k \subseteq \mathbb{K}^\ell \times \mathcal{W}_k = E_k$.

We then apply the two properties of the triangulated category of mixed motives described above, namely the existence of distinguished triangles (2.26) associated to closed embeddings $Y \subset X$ and the homotopy invariance (2.27) inductively to the loci constructed above, all of which are varieties defined over $\mathbb{Q}$. At the first step, single points obviously belong to the category of mixed Tate motives and we are taking the complement $\mathcal{W}_1$ of a point in an affine space, which gives a mixed Tate motive by the first observation above on distinguished triangles associated to closed embeddings. At the next step one considers the complement of the line bundle S_1 inside the trivial vector bundle E_1 over $\mathcal{W}_1$. Again, both $\mathfrak{m}(S_1)$ and $\mathfrak{m}(E_1)$ are mixed Tate motives, since both are products by affine spaces, since we know that the bundles S_k are trivial. Therefore $\mathfrak{m}(E_1 \smallsetminus S_1)$ is also mixed Tate. The same argument shows that, for all $1 \leq k \leq \ell$, the motive $\mathfrak{m}(E_k \smallsetminus S_k)$ is mixed Tate, by repeatedly using the triviality of S_k and the two properties of $\mathcal{DMT}_\mathbb{Q}$ recalled above. $\qquad\square$

The result for the projective hypersurface $\mathcal{D}_\ell$ is then the same, since $\hat{\mathcal{D}}_\ell$ is the affine cone over $\mathcal{D}_\ell$ and the same two properties of the triangulated category of mixed motives used above show that if $\hat{\mathcal{D}}_\ell$ is mixed Tate so is $\mathcal{D}_\ell$. One can use the same geometric construction described above to compute explicitly the class of $\hat{\mathcal{D}}_\ell$ in the Grothendieck ring of varieties. One finds the following expression [Aluffi and Marcolli (2009a)].

Theorem 2.6.3. *The class in the Grothendieck ring of varieties of the affine determinant hypersurface is*

$$[\mathbb{A}^{\ell^2} \smallsetminus \hat{\mathcal{D}}_\ell] = \mathbb{L}^{\binom{\ell}{2}} \prod_{i=1}^{\ell} (\mathbb{L}^i - 1) \tag{2.29}$$

where $\mathbb{L}$ is the class of $\mathbb{A}^1$. In the projective case, the class is

$$[\mathbb{P}^{\ell^2 - 1} \smallsetminus \mathcal{D}_\ell] = \mathbb{L}^{\binom{\ell}{2}} \prod_{i=2}^{\ell} (\mathbb{L}^i - 1). \tag{2.30}$$

Proof. Using the same argument as in Theorem 2.6.2, one shows inductively that the class of $\mathcal{W}_k$ is given by

$$[\mathcal{W}_k] = (\mathbb{L}^\ell - 1)(\mathbb{L}^\ell - \mathbb{L})(\mathbb{L}^\ell - \mathbb{L}^2) \cdots (\mathbb{L}^\ell - \mathbb{L}^{k-1})$$
$$= \mathbb{L}^{\binom{k}{2}}(\mathbb{L}^\ell - 1)(\mathbb{L}^{\ell-1} - 1) \cdots (\mathbb{L}^{\ell-k+1} - 1). \tag{2.31}$$

This suffices to obtain the expression (2.29). The formula (2.30) is then obtained from (2.29) using the fact that $\hat{\mathcal{D}}_\ell$ is the affine cone over $\mathcal{D}_\ell$. □

Notice that, by Theorem 2.6.2, we already know that the class $[\mathbb{A}^{\ell^2} \smallsetminus \hat{\mathcal{D}}_\ell]$ will be in the mixed Tate part $\mathbb{Z}[\mathbb{L}]$ of the Grothendieck ring of varieties $K_0(\mathcal{V}_\mathbb{Q})$. In general, computing the class in the Grothendieck ring may be easier than knowing explicitly the motive in $\mathcal{DM}_\mathbb{Q}$. If one only knows that the class belongs to the Tate part $\mathbb{Z}[\mathbb{L}]$ of the Grothendieck ring, this does not imply that the motive itself will be in the subcategory $\mathcal{DMT}_\mathbb{Q}$ of mixed Tate motives, as in principle there may be cancellations that happen in the Grothendieck ring that do not take place at the level of the motives themselves. However, knowing that the class in the Grothendieck ring is a function of the Lefschetz motive $\mathbb{L}$ is a very good indication of the possible mixed Tate nature of the motive. In fact, as observed for instance in [André (2009)], the Tate conjecture predicts that knowing the number N_p of points of the reduction mod p for almost all p would suffice to determine the motive. For mixed Tate motives the numbers N_p are polynomial in p, and assuming the conjecture holds, knowing that this is the case would conversely imply that the motive is mixed Tate. The role in this of the class in the Grothendieck ring comes from the fact that the counting of points N_p over finite fields has the properties of an Euler characteristic (as we saw above, an additive invariant), hence it factors through the Grothendieck ring. Thus, assuming the conjecture, knowing that the class in the Grothendieck ring is mixed Tate would suffice. Even without these general considerations and the role of the Tate conjecture, in the best cases, the same type of geometric information on stratifications that one can use to compute the class in the Grothendieck ring $K_0(\mathcal{V}_{\mathbb{K}})$ may already contain enough information to run an argument similar to the one used in Theorem 2.6.2, directly at the level of objects in $\mathcal{DM}_{\mathbb{K}}$, as we have seen in the example of the determinant hypersurfaces.

2.7   The algebra of periods

Betti and de Rham cohomology are two of the possible Weil cohomologies for smooth projective varieties. They are related by the *period isomorphism*

$$H_{dR}^*(X) \otimes_{\mathbb{K}} \mathbb{C} \simeq H_B^*(X) \otimes_{\mathbb{Q}} \mathbb{C}, \qquad (2.32)$$

for varieties defined over a number field $\mathbb{K}$. The isomorphism is induced (over affine pieces) by the pairing of differential forms and cycles

$$\omega \otimes \gamma \mapsto \int_\gamma \omega, \qquad (2.33)$$

for

$$\omega \in \mathrm{Ker}(d : \Omega^i(X) \to \Omega^{i+1}(X))$$

and

$$\gamma \in \mathrm{Ker}(\partial_i : C_i(X) \to C_{i-1}(X)),$$

where one identifies Betti cohomology $H_B^*(X) = \mathrm{Hom}(H_*(X), \mathbb{Q}) = H_{sing}^*(X, \mathbb{Q})$ with the dual of singular homology $H_i(X, \mathbb{Z}) = \mathrm{Ker}(\partial_i)/\mathrm{Im}(\partial_{i+1})$. When one is given a basis of $H_{dR}^*(X)$ and of $H_B^*(X)$, the period isomorphism is specified by a *period matrix*.

Periods are typically transcendental numbers. From the number-theoretic point of view they are, arguably, the most interesting class of numbers beyond the algebraic ones. In fact, while being transcendental, they are obtained from a procedure that uses only algebraic data: an algebraic differential form on an algebraic variety, paired with a cycle defined by algebraic relations. Let us denote by $\mathcal{P}_{\mathbb{C}}$ the $\mathbb{Q}$-subalgebra of $\mathbb{C}$ generated by periods. An interesting problem is understanding the structure of this algebra: what relations (over $\mathbb{Q}$) exist between periods.

Following [Kontsevich and Zagier (2001)], one can define a $\mathbb{Q}$-algebra $\mathcal{P}$ as the algebra generated by "formal periods", that is, equivalence classes of elements of the form

$$[(X, D, \omega, \gamma)], \tag{2.34}$$

consisting of a smooth (affine) variety X defined over $\mathbb{Q}$, a normal crossings divisor D in X, an algebraic differential form $\omega \in \Omega^{\dim(X)}(X)$ (not necessarily closed), and a class in the relative homology $\gamma \in H_{\dim(X)}(X(\mathbb{C}), D(\mathbb{C}), \mathbb{Q})$. The equivalence in (2.34) is given by the following relations

- Linearity: the expressions (X, D, ω, γ) are $\mathbb{Q}$-linear in ω and γ.
- Change of variable formula: given a morphism $f : (X, D) \to (X', D')$ and $\omega' \in \Omega^{\dim(X')}(X')$ and $\gamma \in H_{\dim(X)}(X(\mathbb{C}), D(\mathbb{C}), \mathbb{Q})$ one has

$$[(X, D, f^*(\omega), \gamma)] = [(X', D', \omega, f_*(\gamma))]. \tag{2.35}$$

- Stokes' formula:

$$[(X, D, d\omega, \gamma)] = [(\tilde{D}, \tilde{D}^{(1)}, \omega, \partial\gamma)], \tag{2.36}$$

where $\tilde{D}$ is the normalization of D and $\tilde{D}^{(1)}$ is the normalization of $\cup_{i,j}(D_i \cap D_j)$ where D_i are the components of the divisor D.

Kontsevich conjectured that the map $\mathcal{P} \to \mathcal{P}_{\mathbb{C}} \subset \mathbb{C}$ given by

$$[(X, D, \omega, \gamma)] \mapsto \int_{\gamma} \omega \tag{2.37}$$

is injective. This means that all possible non-trivial relations between periods come from either a change of variable formula or Stokes' theorem.

2.8 Mixed Tate motives and the logarithmic extensions

An interesting class of mixed Tate motives is associated to logarithm and polylogarithm functions [Beĭlinson and Deligne (1994)], see also [Ayoub (2007)], [Bloch (1997)].

The extensions $\text{Ext}^1_{\mathcal{DM}(\mathbb{K})}(\mathbb{Q}(0), \mathbb{Q}(1))$ of Tate motives are given by the Kummer motives $M = [\mathbb{Z} \xrightarrow{u} \mathbb{G}_m]$ with $u(1) = q \in \mathbb{K}^*$. This extension has period matrix of the form

$$\begin{pmatrix} 1 & 0 \\ \log q & 2\pi i \end{pmatrix}. \tag{2.38}$$

When, instead of working with motives over the base field $\mathbb{K}$, one works with the relative setting of motivic sheaves over a base scheme S, instead of the Tate motives $\mathbb{Q}(n)$ one considers the Tate sheaves $\mathbb{Q}_S(n)$. These correspond to the constant sheaf with the motive $\mathbb{Q}(n)$ over each point $s \in S$. In the case where $S = \mathbb{G}_m$, there is a natural way to assemble the Kummer motives into a unique extension in $\text{Ext}^1_{\mathcal{DM}(\mathbb{G}_m)}(\mathbb{Q}_{\mathbb{G}_m}(0), \mathbb{Q}_{\mathbb{G}_m}(1))$. This is the Kummer extension

$$\mathbb{Q}_{\mathbb{G}_m}(1) \to \mathcal{K} \to \mathbb{Q}_{\mathbb{G}_m}(0) \to \mathbb{Q}_{\mathbb{G}_m}(1)[1], \tag{2.39}$$

where over the point $s \in \mathbb{G}_m$ one is taking the Kummer extension $M_s = [\mathbb{Z} \xrightarrow{u} \mathbb{G}_m]$ with $u(1) = s$. Because of the logarithm function $\log(s)$ that appears in the period matrix for this extension, the Kummer extension (2.39) is also referred to as the *logarithmic motive*, and denoted by $\mathcal{K} = \text{Log}$, *cf.* [Ayoub (2007)] [Beǐlinson and Deligne (1994)].

Let $\mathcal{DM}(\mathbb{G}_m)$ be the Voevodsky category of mixed motives (motivic sheaves) over the multiplicative group $\mathbb{G}_m$. When working with $\mathbb{Q}$-coefficients, so that one can include denominators in the definition of projectors, one can then consider the logarithmic motives Log^n, defined by

$$\text{Log}^n = \text{Sym}^n(\mathcal{K}), \tag{2.40}$$

where the symmetric powers of an object in $\mathcal{DM}_{\mathbb{Q}}(\mathbb{G}_m)$ are defined as

$$\text{Sym}^n(X) = \frac{1}{\#\Sigma_n} \sum_{\sigma \in \Sigma_n} \sigma(X^n). \tag{2.41}$$

The polylogarithms appear naturally as period matrices for extensions involving the symmetric powers $\text{Log}^n = \text{Sym}^n(\mathcal{K})$, in the form [Bloch (1997)]

$$0 \to \text{Log}^{n-1}(1) \to \mathcal{L}^n \to \mathbb{Q}(0) \to 0, \tag{2.42}$$

where $M(1) = M \otimes \mathbb{Q}(1)$ and $\mathcal{L}^1 = \text{Log}$. The mixed motive $\mathcal{L}^n$ has period matrix

$$\begin{pmatrix} 1 & 0 \\ M_{\text{Li}}^{(n)} & M_{\text{Log}^{n-1}(1)} \end{pmatrix} \tag{2.43}$$

with

$$M_{\text{Li}}^{(n)} = (-\text{Li}_1(s), -\text{Li}_2(s), \cdots, -\text{Li}_n(s))^\tau, \qquad (2.44)$$

where τ means transpose and where

$$\text{Li}_1(s) = -\log(1-s), \quad \text{and} \quad \text{Li}_n(s) = \int_0^s \text{Li}_{n-1}(u) \frac{du}{u},$$

equivalently defined (on the principal branch) using the power series

$$\text{Li}_n(s) = \sum_k \frac{s^k}{k^n},$$

and with

$$M_{\text{Log}^n(1)} = \begin{pmatrix} 2\pi i & 0 & 0 & \cdots & 0 \\ 2\pi i \log(s) & (2\pi i)^2 & 0 & \cdots & 0 \\ 2\pi i \frac{\log^2(s)}{2!} & (2\pi i)^2 \log(s) & (2\pi i)^3 & \cdots & 0 \\ \vdots & \vdots & \vdots & \cdots & \vdots \\ 2\pi i \frac{\log^n(s)}{n!} & (2\pi i)^2 \frac{\log^{n-1}(s)}{(n-1)!} & (2\pi i)^3 \frac{\log^{n-2}(s)}{(n-2)!} & \cdots & (2\pi i)^n \end{pmatrix}.$$

$$(2.45)$$

The period matrices for the motives Log^n correspond to the description of Log^n as the extension of $\mathbb{Q}(0)$ by $\text{Log}^{n-1}(1)$, *i.e.* to the distinguished triangles in $\mathcal{DM}(\mathbb{G}_m)$ of the form

$$\text{Log}^{n-1}(1) \to \text{Log}^n \to \mathbb{Q}(0) \to \text{Log}^{n-1}(1)[1]. \qquad (2.46)$$

The motives Log^n form a projective system under the canonical maps

$$\beta_n : \text{Log}^{n+1} \to \text{Log}^n$$

given by the composition of the morphisms $\text{Sym}^{n+m}(\mathcal{K}) \to \text{Sym}^n(\mathcal{K}) \otimes \text{Sym}^m(\mathcal{K})$, as in [Ayoub (2007)], Lemma 4.35, given by the fact that $\text{Sym}^{n+m}(\mathcal{K})$ is canonically a direct factor of $\text{Sym}^n(\mathcal{K}) \otimes \text{Sym}^m(\mathcal{K})$, and the map $\text{Sym}^m(\mathcal{K}) \to \mathbb{Q}(0)$ of (2.46), in the particular case $m = 1$. Let Log^∞ denote the pro-motive obtained as the projective limit

$$\text{Log}^\infty = \varprojlim_n \text{Log}^n. \qquad (2.47)$$

The analog of the period matrix (2.45) then becomes the infinite matrix

$$M_{\text{Log}^\infty(1)} = \begin{pmatrix} 2\pi i & 0 & 0 & \cdots & 0 & \cdots \\ 2\pi i \log(s) & (2\pi i)^2 & 0 & \cdots & 0 & \cdots \\ 2\pi i \frac{\log^2(s)}{2!} & (2\pi i)^2 \log(s) & (2\pi i)^3 & \cdots & 0 & \cdots \\ \vdots & \vdots & \vdots & \cdots & \vdots & \cdots \\ 2\pi i \frac{\log^n(s)}{n!} & (2\pi i)^2 \frac{\log^{n-1}(s)}{(n-1)!} & (2\pi i)^3 \frac{\log^{n-2}(s)}{(n-2)!} & \cdots & (2\pi i)^n & \cdots \\ \vdots & \vdots & \vdots & \cdots & \vdots & \cdots \end{pmatrix}.$$

$$(2.48)$$

This can best be expressed in terms of *mixed Hodge structures*. These comprise the analytic side of the theory of motives and they provide one of the possible realizations of categories of mixed motives, hence another way to test complexity of various types of mixed motives. One has the following definition as in [Steenbrink (1994)].

Definition 2.8.1. A $\mathbb{Q}$-mixed Hodge structure consists of the data $M = (V, W_\bullet, F^\bullet)$ of a finite-dimensional $\mathbb{Q}$-vector space, with an increasing filtration $W_\bullet V$ (weight filtration) and a decreasing filtration $F^\bullet V_\mathbb{C}$ for $V_C = V \otimes_\mathbb{Q} \mathbb{C}$ (the Hodge filtration), satisfying

$$\mathrm{gr}_n^W V_\mathbb{C} = F^p \mathrm{gr}_n^W V_\mathbb{C} \oplus \overline{F^{n-p+1}} \mathrm{gr}_n^W V_\mathbb{C}, \qquad (2.49)$$

where $\bar{F}^\bullet$ is the complex conjugate of $F^\bullet$ with respect to $V_\mathbb{R} \subset V_\mathbb{C}$.

The Hodge structure $(V, W_\bullet, F^\bullet)$ is pure of weight n if $W_n = V$ and $W_{n-1} = 0$.

A morphism $M_1 \to M_2$ of $\mathbb{Q}$-MHS is a linear map $\phi : V_1 \to V_2$ such that $\phi(W_k V_1) \subset W_k V_2$ and $\phi(F^p V_{1,\mathbb{C}}) \subset F^p V_{2,\mathbb{C}}$. Tensoring of $\mathbb{Q}$-MHS is also defined, by considering $V_1 \otimes_\mathbb{Q} V_2$ with filtrations

$$W_m(V_1 \otimes_\mathbb{Q} V_2) = \sum_{i+j=m} W_i(V_1) \otimes_\mathbb{Q} W_j(V_2),$$

$$F^p(V_1 \otimes_\mathbb{Q} V_2) = \sum_{r+s=p} F^r(V_{1,\mathbb{C}}) \otimes_\mathbb{C} F^s(V_{2,\mathbb{C}}).$$

A $\mathbb{Q}$-mixed Hodge structure $M = (V, W_\bullet, F^\bullet)$ is *mixed Tate* if

$$\mathrm{gr}_{2p}^W M = \oplus \mathbb{Q}(-p), \quad \text{and} \quad \mathrm{gr}_{2p-1}^W M = 0.$$

The condition (2.49) is equivalent to a Hodge decomposition

$$\mathrm{gr}_n^W V_\mathbb{C} = \oplus_{p+q=n} V^{p,q}, \qquad (2.50)$$

$$V^{p,q} = F^p \mathrm{gr}_{p+q}^W V_\mathbb{C} \cap \overline{F^q} \mathrm{gr}_{p+q}^W V_\mathbb{C}. \qquad (2.51)$$

An example of pure Hodge structure of weight $-2n$ is the Tate Hodge structure with $V = (2\pi i)^n \mathbb{Q}$. These correspond to the pure Tate motives. These define, by tensoring, *Tate twists* on $\mathbb{Q}$-mixed Hodge structures.

For Hodge–Tate structures one has an associated *period matrix* obtained as in §2 of [Goncharov (1999)], as the composition $S_{HT}^{-1} \circ S_W$ of an isomorphism $S_{HT} : \oplus_p \mathrm{gr}_{2p}^W(V_\mathbb{Q}) \otimes_\mathbb{Q} \mathbb{C} \to V_\mathbb{C}$ and a splitting of the weight filtration, which is an isomorphism $S_W : \oplus_p \mathrm{gr}_{2p}^W(V_\mathbb{Q}) \otimes_\mathbb{Q} \mathbb{C} \to V_\mathbb{C}$, expressed in a chosen basis of $\mathrm{gr}_{2p}^W V_\mathbb{Q}$.

In the example above, the mixed Hodge structure associated to the motives Log^n is the one that has as the weight filtrations W_{-2k} the range of multiplication by the matrix M_{Log^n} defined as in (2.45) on vectors in $\mathbb{Q}^n$ with the first $k - 1$ entries equal to zero, while the Hodge filtration F^{-k} is given by the range of multiplication of M_{Log^n} on vectors of $\mathbb{C}^n$ with the entries from $k + 1$ to n equal to zero [Bloch (1997)].

2.9 Categories and Galois groups

We review here a type of structure on a category which is more rigid than that of an abelian category we encountered before, and which is sufficient to identify it uniquely as a category of finite dimensional linear representations of a group, namely the notion of *Tannakian category*, see [Saavedra Rivano (1972)], [Deligne and Milne (1982)].

Definition 2.9.1. A k-linear rigid abelian tensor category $\mathcal{C}$ is a Tannakian category if it admits a fiber functor, *i.e.* an exact faithful tensor functor, $\omega : \mathcal{C} \to \text{Vect}_K$, with target the category of finite-dimensional vector spaces over a field extension K of k. A Tannakian category $\mathcal{C}$ is neutral if $K = k$.

Recall here that a functor $\omega : \mathcal{C} \to \mathcal{C}'$ is faithful if, for all $X, Y \in \text{Obj}(\mathcal{C})$, the mapping $\omega : \text{Hom}_{\mathcal{C}}(X, Y) \to \text{Hom}_{\mathcal{C}'}(\omega(X), \omega(Y))$ is injective. If $\mathcal{C}$ and $\mathcal{C}'$ are k-linear categories, a functor ω is additive if the map of Hom sets above is a k-linear map. An additive functor ω is exact if, for any exact sequence $0 \to X \to Y \to Z \to 0$ in $\mathcal{C}$, the corresponding sequence $0 \to \omega(X) \to \omega(Y) \to \omega(Z) \to 0$ in $\mathcal{C}'$ is also exact. A functor $\omega : \mathcal{C} \to \mathcal{C}'$ between k-linear tensor categories is a tensor functor if there are functorial isomorphisms $\tau_1 : \omega(1) \to 1$ and $\tau_{X,Y} : \omega(X \otimes Y) \to \omega(X) \otimes \omega(Y)$.

The main result of the theory of Tannakian categories, of which we are going to see an application to motives in the next section and one to quantum field theory in a later chapter, is the following statement that reconstructs a group from the category.

Theorem 2.9.2. [Saavedra Rivano (1972)] *Let $\mathcal{C}$ be a neutral Tannakian category, with fiber functor $\omega : \mathcal{C} \to \text{Vect}_k$. Let G be the group of automorphisms (invertible natural transformations) of the fiber functor. Then ω induces an equivalence of rigid tensor categories $\omega : \mathcal{C} \to \text{Rep}_G$, where Rep_G is the category of finite dimensional linear representations of G.*

In fact, here G is really an affine group scheme in the sense that we are going to recall in §5.2 below. The affine group scheme G is referred to as the Tannakian Galois group of $\mathcal{C}$.

If a Tannakian category $\mathcal{C}$ is semi-simple, then it is known that the Tannakian Galois group is reductive. For example, the Tannakian Galois group of the category of pure motives is reductive. So is the one of the Tannakian subcategory generated by the pure Tate motives $\mathbb{Q}(n)$. In fact, the latter is the simplest example of a reductive group, the multiplicative group $\mathbb{G}_m$.

2.10   Motivic Galois groups

We recalled above that, over a number field $\mathbb{K}$, it is possible to form an abelian category $\mathcal{MT}(\mathbb{K})$ of mixed Tate motives, as the heart of a t-structure inside the triangulated subcategory $\mathcal{DMT}(\mathbb{K})$ of the triangulated category $\mathcal{DM}(\mathbb{K})$ of mixed motives. In fact, over a number field one has an explicit description of the extensions in $\mathcal{DM}(\mathbb{K})$ between pure Tate motives in terms of algebraic K-theory as

$$\mathrm{Ext}^1_{\mathcal{DM}(\mathbb{K})}(\mathbb{Q}(0), \mathbb{Q}(n)) = K_{2n-1}(\mathbb{K}) \otimes \mathbb{Q} \qquad (2.52)$$

and $\mathrm{Ext}^2_{\mathcal{DM}(\mathbb{K})}(\mathbb{Q}(0), \mathbb{Q}(n)) = 0$, see [Levine (1993)] and [Levine (1998)], Appendix B. For example, the mixed Tate motives that correspond to extensions in $\mathrm{Ext}^1_{\mathcal{DM}(\mathbb{K})}(\mathbb{Q}(0), \mathbb{Q}(1))$ are the Kummer motives we have encountered in the previous sections.

The resulting abelian category $\mathcal{MT}(\mathbb{K})$ of mixed Tate motives over a number field is in fact also a Tannakian category, with fiber functor ω to $\mathbb{Z}$-graded $\mathbb{Q}$-vector spaces, $M \mapsto \omega(M) = \oplus_n \omega_n(M)$ with

$$\omega_n(M) = \mathrm{Hom}(\mathbb{Q}(n), \mathrm{Gr}^w_{-2n}(M)), \qquad (2.53)$$

where $\mathrm{Gr}^w_{-2n}(M) = W_{-2n}(M)/W_{-2(n+1)}(M)$ is the graded structure associated to the finite increasing weight filtration W.

The construction of a Tannakian category of mixed Tate motives $\mathcal{MT}(\mathcal{O}_V)$ and the explicit form of its motivic Galois group are also known for motivic sheaves over the scheme $\mathcal{O}_V$ of V-integers of a number field $\mathbb{K}$, for V a set of finite places of $\mathbb{K}$. This means that objects in this category are mixed Tate motives over $\mathbb{K}$ that are unramified at each finite place $v \notin V$. In this case one also has a K-theory group parameterizing extensions in the form

$$\mathrm{Ext}^1_{\mathcal{DM}(\mathcal{O}_V)}(\mathbb{Q}(0), \mathbb{Q}(n)) = \begin{cases} K_{2n-1}(\mathbb{K}) \otimes \mathbb{Q} & n \geq 2 \\ \mathcal{O}_V^* \otimes \mathbb{Q} & n = 1 \\ 0 & n \leq 0. \end{cases} \tag{2.54}$$

and $\mathrm{Ext}^2_{\mathcal{DM}(\mathcal{O}_V)}(\mathbb{Q}(0), \mathbb{Q}(n)) = 0$. The affine group scheme that gives the motivic Galois group in this case is of the form $\mathbb{U} \rtimes \mathbb{G}_m$, where the reductive group $\mathbb{G}_m$ comes from the pure Tate motives, and $\mathbb{U}$ is a pro-unipotent affine group scheme, that takes into account the nontrivial extensions between pure Tate motives. In fact, the corresponding Lie algebra $\mathrm{Lie}(\mathbb{U})$ is freely generated by a set of homogeneous generators in degree n identified with a basis of the dual of $\mathrm{Ext}^1(\mathbb{Q}(0), \mathbb{Q}(n))$, as proved in Proposition 2.3 of [Deligne and Goncharov (2005)]. There is however *no canonical identification* between $\mathrm{Lie}(\mathbb{U})$ and the free Lie algebra generated by the graded vector space $\oplus \mathrm{Ext}^1(\mathbb{Q}(0), \mathbb{Q}(n))^\vee$.

The following special case of this general result of [Deligne and Goncharov (2005)] is especially relevant to the context of renormalization, since it provides a motivic Galois group (non-canonically) isomorphic to the Tannakian Galois group of renormalization of [Connes and Marcolli (2004)], which we review in Chapter 6.

Proposition 2.10.1. [Deligne and Goncharov (2005)] *For* $\mathbb{K} = \mathbb{Q}(\zeta_N)$ *the cyclotomic field of level* $N = 3$ *or* $N = 4$ *and* $\mathcal{O}$ *its ring of integers, the motivic Galois group of the Tannakian category* $\mathcal{MT}(\mathcal{O}[1/N])$ *is of the form* $\mathbb{U} \rtimes \mathbb{G}_m$, *where the Lie algebra* $\mathrm{Lie}(\mathbb{U})$ *is (non-canonically) isomorphic to the free graded Lie algebra with one generator* e_n *in each degree* $n \leq -1$.

Chapter 3

Feynman integrals and algebraic varieties

In this chapter we begin to connect the two topics introduced above, perturbative quantum field theory and motives. We follow first what we referred to as the "bottom-up approach" in the introduction, by showing how to associate to individual Feynman integrals an algebraic variety, in the form of a hypersurface complement, and a (possibly divergent) period integral on a chain with boundary on a normal crossings divisor. The motivic nature of this period, and the question of whether this (after removing divergences) will be a period of a mixed Tate motive can then be formulated as the question of whether a certain relative cohomology of the hypersurface complement relative to the normal crossings divisor is a realization of a mixed Tate motive. We begin by describing the parametric form of Feynman integrals and the main properties of the two Kirchhoff–Symanzik graph polynomials involved in this formulation and the graph hypersurfaces defined by the first polynomial. We show, as in [Aluffi and Marcolli (2008a)], how one can sometimes obtain explicit computations of the classes in the Grothendieck ring of varieties for some families of Feynman graphs, using the classical Cremona transformation and dual graphs. These classes also satisfy a form of deletion/contraction relation, as proved in [Aluffi and Marcolli (2009b)]. We then describe, following the results of [Aluffi and Marcolli (2008b)], a way of formulating Feynman rules directly at the algebro-geometric level of graph hypersurface complements. We give examples of such Feynman rules that factor through the Grothendieck ring and others, defined using characteristic classes of singular varieties, that do not descend to the level of isomorphism classes of varieties in the Grothendieck ring, although they still satisfy an inclusion-exclusion principle. We discuss the difference between working with affine or projective hypersurfaces in describing Feynman integrals. Finally, following [Aluffi and Marcolli (2009a)], we show that one can reformulate the period computation underlying the Feynman amplitudes in terms of a simpler hypersurface complement, using determinant hypersurfaces, whose mixed Tate nature is established, but with a more complicated normal crossings divisor as the second term in the relative cohomology whose motivic nature one wishes to establish. We describe the advantages and the difficulties of this viewpoint.

3.1   The parametric Feynman integrals

Using the Feynman parameters introduced in §1.7 above, we show how to reformulate the Feynman integral

$$U(\Gamma, p_1, \ldots, p_N) = \int \frac{\delta(\sum_{i=1}^n \epsilon_{v,i} k_i + \sum_{j=1}^N \epsilon_{v,j} p_j)}{q_1(k_1) \cdots q_n(k_n)} \, d^D k_1 \cdots d^D k_n$$

in the form known as the *Feynman parametric representation*. (We neglect here an overall multiplicative factor in the coupling constants and powers of 2π.)

The first step is to rewrite the denominator $q_1 \cdots q_n$ of (1.37) in the form of an integration over the topological simplex σ_n as in (1.51), in terms of the Feynman parameters $t = (t_1, \ldots, t_n) \in \sigma_n$.

In writing the integral (1.37) we have made a choice of an orientation of the graph Γ, since the matrix $\epsilon_{v,i}$ involved in writing the conservation laws at vertices in (1.37) depends on the orientation given to the edges of the graph. Now we also make a choice of a set of generators for the first homology group $H_1(\Gamma, \mathbb{Z})$, *i.e.* a choice of a maximal set of independent loops in the graph, $\{l_1, \ldots, l_\ell\}$ with $\ell = b_1(\Gamma)$ the first Betti number.

We then define another matrix associated to the graph Γ, the *circuit matrix* $\eta = (\eta_{ik})$, with $i \in E(\Gamma)$ and $k = 1, \ldots, \ell$ ranging over the chosen basis of loops, given by

$$\eta_{ik} = \begin{cases} +1 & \text{if edge } e_i \in \text{ loop } l_k, \text{ same orientation} \\ -1 & \text{if edge } e_i \in \text{ loop } l_k, \text{ reverse orientation} \\ 0 & \text{if edge } e_i \notin \text{ loop } l_k. \end{cases} \qquad (3.1)$$

There is a relation between the circuit matrix and the incidence matrix of the graph, which is given as follows.

Lemma 3.1.1. *The incidence matrix $\epsilon = (\epsilon_{v,i})$ and the circuit matrix $\eta = (\eta_{ik})$ of a graph Γ satisfy the relation $\epsilon \eta = 0$. This holds independently of the choice of the orientation of the graph and the basis of $H_1(\Gamma, \mathbb{Z})$.*

Proof. The result follows from the observation that, independently of the choices of orientations, for two given edges e_i and e_j in a loop l_k, both incident to a vertex v, one has

$$\epsilon_{v,i}\eta_{ik} = -\epsilon_{v,j}\eta_{jk}$$

so that one obtains

$$\sum_i \epsilon_{v,i}\eta_{ik} = 0,$$

since $\epsilon_{v,i}\eta_{ik} \neq 0$ for only two edges in the loop l_k, with different signs, as above; see §2.2 of [Nakanishi (1971)]. $\square$

We then define the Kirchhoff matrix of the graph, also known as the Symanzik matrix.

Definition 3.1.2. The Kirchhoff–Symanzik matrix $M_\Gamma(t)$ of the graph Γ is the $\ell \times \ell$ matrix given by

$$(M_\Gamma(t))_{kr} = \sum_{i=1}^n t_i \eta_{ik} \eta_{ir}. \tag{3.2}$$

Equivalently, it can be written as

$$M_\Gamma(t) = \eta^\dagger \Lambda(t) \eta,$$

where $\dagger$ is the transpose and $\Lambda(t)$ is the diagonal matrix with entries $(t_1, \ldots, t_n)$. We think of M_Γ as a function

$$M_\Gamma : \mathbb{A}^n \to \mathbb{A}^{\ell^2}, \quad t = (t_1, \ldots, t_n) \mapsto M_\Gamma(t) = (M_\Gamma(t))_{kr} \tag{3.3}$$

where $\mathbb{A}$ denotes the affine line over a field (here mostly $\mathbb{C}$ or $\mathbb{R}$ or $\mathbb{Q}$).

Definition 3.1.3. The Kirchhoff–Symanzik polynomial $\Psi_\Gamma(t)$ of the graph Γ is defined as

$$\Psi_\Gamma(t) = \det(M_\Gamma(t)). \tag{3.4}$$

Notice that, while the construction of the matrix $M_\Gamma(t)$ depends on the choice of an orientation on the graph Γ and of a basis of $H_1(\Gamma, \mathbb{Z})$, the graph polynomial is independent of these choices.

Lemma 3.1.4. *The Kirchhoff–Symanzik polynomial $\Psi_\Gamma(t)$ is independent of the choice of edge orientation and of the choice of generators for $H_1(\Gamma, \mathbb{Z})$.*

Proof. A change of orientation in a given edge results in a change of sign in one of the columns of $\eta = \eta_{ik}$. The change of sign in the corresponding row of $\eta^\dagger$ leaves the determinant of $M_\Gamma(t) = \eta^\dagger \Lambda(t)\eta$ unaffected. A change of basis for $H_1(\Gamma, \mathbb{Z})$ changes $M_\Gamma(t) \mapsto AM_\Gamma(t)A^{-1}$, where $A \in \mathrm{GL}_\ell(\mathbb{Z})$ is the matrix that gives the change of basis. The determinant is once again unchanged. $\qquad\square$

We view it as a function $\Psi_\Gamma : \mathbb{A}^n \to \mathbb{A}$. We define the affine graph hypersurface $\hat{X}_\Gamma$ to be the locus of zeros of the graph polynomial
$$\hat{X}_\Gamma = \{t \in \mathbb{A}^n \mid \Psi_\Gamma(t) = 0\}. \tag{3.5}$$
The polynomial Ψ_Γ is by construction a homogeneous polynomial of degree $\ell = b_1(\Gamma)$, hence we can view it as defining a hypersurface in projective space $\mathbb{P}^{n-1} = (\mathbb{A}^n \smallsetminus \{0\})/\mathbb{G}_m$,
$$X_\Gamma = \{t \in \mathbb{P}^{n-1} \mid \Psi_\Gamma(t) = 0\}, \tag{3.6}$$
of which $\hat{X}_\Gamma$ is the affine cone $\hat{X}_\Gamma = C(X_\Gamma)$.

After rewriting the denominator of the integrand in (1.37) in terms of an integration over σ_n using the Feynman parameters, we want to replace in the Feynman integral $U(\Gamma, p_1, \ldots, p_N)$ the variables k_i associated to the internal edges, and the integration in these variables, by variables x_r associated to the independent loops in the graph and an integration only over these variables, using the linear constraints at the vertices. We set
$$k_i = u_i + \sum_{r=1}^{\ell} \eta_{ir} x_r, \tag{3.7}$$
with the constraint
$$\sum_{i=1}^{n} t_i u_i \eta_{ir} = 0, \quad \forall r = 1, \ldots, \ell, \tag{3.8}$$
that is, we require that the column vector $\Lambda(t)u$ is orthogonal to the rows of the circuit matrix η.

The momentum conservation conditions in the delta function in the numerator of (1.37) give
$$\sum_{i=1}^{n} \epsilon_{v,i} k_i + \sum_{j=1}^{N} \epsilon_{v,j} p_j = 0. \tag{3.9}$$

Lemma 3.1.5. *Using the change of variables* (3.7) *and the constraint* (3.8) *one finds the conservation condition*
$$\sum_{i=1}^{n} \epsilon_{v,i} u_i + \sum_{j=1}^{N} \epsilon_{v,j} p_j = 0. \tag{3.10}$$

Proof. This follows immediately from the orthogonality relation between the incidence matrix and circuit matrix of Lemma 3.1.1. □

The two equations (3.8) and (3.10) constitute the Krichhoff laws of circuits applied to the flow of momentum through the Feynman graph. In particular they determine uniquely the $u_i = u_i(p)$ as functions of the external momenta. We see the explicit form of the solution in Proposition 3.1.7 below. First we give a convenient reformulation of the graph polynomial.

The graph polynomial $\Psi_\Gamma(t)$ has a more explicit combinatorial description in terms of the graph Γ, as follows.

Proposition 3.1.6. *The Kirchhoff–Symanzik polynomial $\Psi_\Gamma(t)$ of (3.4) is given by*

$$\Psi_\Gamma(t) = \sum_{T \subset \Gamma} \prod_{e \notin E(T)} t_e, \qquad (3.11)$$

where the sum is over all the spanning trees T of the graph Γ and for each spanning tree the product is over all edges of Γ that are not in that spanning tree.

Proof. The polynomial $\Psi_\Gamma = \det(M_\Gamma(t))$ can equivalently be described as the polynomial (see [Itzykson and Zuber (2006)], §6-2-3 and [Nakanishi (1971)] §1.3-2)

$$\Psi_\Gamma(t) = \sum_S \prod_{e \in S} t_e, \qquad (3.12)$$

where S ranges over all the subsets $S \subset E_{int}(\Gamma)$ of $\ell = b_1(\Gamma)$ internal edges of Γ, such that the removal of all the edges in S leaves a connected graph. This can be seen to be equivalent to the formulation (3.11) in terms of spanning trees of the graph Γ (see [Nakanishi (1971)] §1.3). In fact, each spanning tree has $\#V(\Gamma) - 1$ edges and is the complement of a set S as above. □

In the case of graphs with several connected components, one defines the Kirchhoff–Symanzik polynomial as in (3.11), with the sum over *spanning forests*, by which we mean here collections of a spanning tree in each connected component. This polynomial is therefore multiplicative over connected components.

We then have the following description of the resulting term $\sum_i t_i u_i^2$ after the change of variables (3.7). This will be useful in Theorem 3.1.9 below.

Proposition 3.1.7. *The term* $\sum_i t_i u_i^2$ *is of the form* $\sum_i t_i u_i^2 = p^\dagger R_\Gamma(t) p$, *where* $R_\Gamma(t)$ *is an* $N \times N$ *matrix, with* $N = \#E_{ext}(\Gamma)$ *with*

$$p^\dagger R_\Gamma(t) p = \sum_{v,v' \in V(\Gamma)} P_v (D_\Gamma(t)^{-1})_{v,v'} P_{v'},$$

with

$$(D_\Gamma(t))_{v,v'} = \sum_{i=1}^n \epsilon_{v,i} \epsilon_{v',i} t_i^{-1} \tag{3.13}$$

and

$$P_v = \sum_{e \in E_{ext}(\Gamma), t(e)=v} p_e. \tag{3.14}$$

Proof. We give a quick sketch of the argument and we refer the reader to [Nakanishi (1971)] and [Itzykson and Zuber (2006)], §6-2-3 for more details. The result is a consequence of the fact that the u_i, as functions of the external momenta $p = (p_j)$, are determined by the Kirchhoff law (3.10). Thus, there is a matrix $A_{i,v}$ such that $-\sum_v A_{i,v} P_v = t_i u_i$, hence $\sum_i t_i u_i^2 = \sum_i \sum_{v,v'} P_v P_{v'} A_{i,v} A_{i,v'} t_i^{-1}$. The constraints on the u_i given by the Kirchhoff laws also show that $A_{i,v} = \epsilon_{i,v}$ so that one obtains the matrix $D_\Gamma(t)$ as in (3.13). $\qquad\square$

We set

$$V_\Gamma(t,p) = p^\dagger R_\Gamma(t) p + m^2. \tag{3.15}$$

In the massless case $(m = 0)$, we will see below that this is a ratio of two homogeneous polynomials in t,

$$V_\Gamma(t,p)|_{m=0} = \frac{P_\Gamma(t,p)}{\Psi_\Gamma(t,p)}, \tag{3.16}$$

of which the denominator is the graph polynomial (3.4) and $P_\Gamma(t,p)$ is a homogeneous polynomial of degree $b_1(\Gamma) + 1$. We have the following result; see [Itzykson and Zuber (2006)], §6-2-3.

Proposition 3.1.8. *The function* $V_\Gamma(t,p)$ *satisfies* (3.16) *in the massless case, with the polynomial* $P_\Gamma(t,p)$ *of the form*

$$P_\Gamma(t,p) = \sum_{C \subset \Gamma} s_C \prod_{e \in C} t_e, \tag{3.17}$$

where the sum is over the cut-sets $C \subset \Gamma$, *i.e. the collections of* $b_1(\Gamma) + 1$ *edges that divide the graph* Γ *into exactly two connected components* $\Gamma_1 \cup \Gamma_2$.

The coefficient s_C is a function of the external momenta attached to the vertices in either one of the two components,

$$s_C = \left(\sum_{v \in V(\Gamma_1)} P_v \right)^2 = \left(\sum_{v \in V(\Gamma_2)} P_v \right)^2, \qquad (3.18)$$

where the P_v are defined as in (3.14), as the sum of the incoming external momenta.

Proof. We give a brief sketch of the argument and refer the reader to the more detailed treatment in [Nakanishi (1971)]. The matrix $D_\Gamma(t)$ of (3.13) has determinant

$$\det(D_\Gamma(t)) = \sum_T \prod_{e \in T} t_e^{-1},$$

where T ranges over the spanning trees of the graph Γ. This is related to the polynomial $\Psi_\Gamma(t)$ by

$$\Psi_\Gamma(t) = (\prod_e t_e) \det(D_\Gamma(t)) = \sum_T \prod_{e \notin T} t_e.$$

Thus, for $m = 0$, the function (3.15) becomes

$$p^\dagger R_\Gamma(t) p = \sum_{v,v' \in V(\Gamma)} P_v (D_\Gamma(t)^{-1})_{v,v'} P_{v'}$$

$$= \frac{1}{\Psi_\Gamma(t)} \sum_{C \subset \Gamma} s_C \prod_{e \in C} t_e. \qquad \square$$

We can now rewrite the Feynman integral in its parametric form as follows; see [Bjorken and Drell (1964)] §8 and [Bjorken and Drell (1965)] §18.

Theorem 3.1.9. *Up to a multiplicative constant $C_{n,\ell}$, the Feynman integral $U(\Gamma, p_1, \ldots, p_N)$ can be equivalently written in the form*

$$U(\Gamma, p_1, \ldots, p_N) = \frac{\Gamma(n - \frac{D\ell}{2})}{(4\pi)^{D\ell/2}} \int_{\sigma_n} \frac{\omega_n}{\Psi_\Gamma(t)^{D/2} V_\Gamma(t,p)^{n - D\ell/2}}, \qquad (3.19)$$

where ω_n is the volume form on the simplex σ_n.

Proof. We first show that we have

$$\int \frac{d^D x_1 \cdots d^D x_\ell}{(\sum_{i=0}^n t_i q_i)^n} = C_{\ell,n} \det(M_\Gamma(t))^{-D/2} (\sum_{i=0}^n t_i(u_i^2 + m^2))^{-n + D\ell/2}, \qquad (3.20)$$

where $u_i = u_i(p)$ as above. In fact, after the change of variables (3.7), the left hand side of (3.20) reads

$$\int \frac{d^D x_1 \cdots d^D x_\ell}{(\sum_{i=0}^n t_i(u_i^2 + m^2) + \sum_{k,r}(M_\Gamma)_{kr} x_k x_r)^n}.$$

The integral can then be reduced by a further change of variables that diagonalizes the matrix M_Γ to an integral of the form

$$\int \frac{d^D y_1 \cdots d^D y_\ell}{(a + \sum_k \lambda_k y_k^2)^n} = C_{\ell,n} \, a^{-n+D\ell/2} \prod_{k=1}^{\ell} \lambda_k^{-D/2},$$

with

$$C_{\ell,n} = \int \frac{d^D x_1 \cdots d^D x_\ell}{(1 + \sum_k x_k^2)^n}.$$

We then write $\det M_\Gamma(t) = \Psi_\Gamma(t)$ and we use the expression of Proposition 3.1.7 to express the term $(\sum_i t_i(u_i^2 + m^2))^{-n+D\ell/2}$ in terms of

$$\sum_i t_i(u_i^2 + m^2) = \sum_i t_i u_i^2 + m^2 = V_\Gamma(t, p),$$

with $V_\Gamma(t, p)$ as in (3.15).   □

In the *massless* case, using the expression (3.16) of $V_\Gamma(t, p)$ in terms of the polynomials $P_\Gamma(t, p)$ and $\Psi_\Gamma(t)$, one writes equivalently the integral $U(\Gamma, p)$ as

$$U(\Gamma, p_1, \ldots, p_N) = \frac{\Gamma(n - \frac{D\ell}{2})}{(4\pi)^{D\ell/2}} \int_{\sigma_n} \frac{P_\Gamma(t, p)^{-n+D\ell/2} \omega_n}{\Psi_\Gamma(t)^{-n+D(\ell+1)/2}}. \tag{3.21}$$

Up to a divergent Gamma-factor, one is interested in understanding the motivic nature (*i.e.* the nature as a period) of the remaining integral

$$\mathcal{I}(\Gamma, p_1, \ldots, p_N) = \int_{\sigma_n} \frac{P_\Gamma(t, p)^{-n+D\ell/2} \omega_n}{\Psi_\Gamma(t)^{-n+D(\ell+1)/2}}. \tag{3.22}$$

3.2 The graph hypersurfaces

We introduced above the projective hypersurfaces $X_\Gamma \subset \mathbb{P}^{n-1}$, with $n = \#E_{int}(\Gamma)$, and the corresponding affine hypersurface $\hat{X}_\Gamma \subset \mathbb{A}^n$.

We observe here that these hypersurfaces are in general *singular* and with singular locus of positive dimension; see §3.5 below for more details. The fact that we are dealing with singular hypersurfaces in projective spaces implies that, when we consider motives associated to these geometric objects, we will necessarily be dealing with *mixed* motives.

We consider only the case where the spacetime dimension D of the quantum field theory is *sufficiently large*. This means that we look at the parametric Feynman integrals (3.19) in the *stable range* where

$$n \leq D\ell/2, \tag{3.23}$$

for $n = \#E_{int}(\Gamma)$ and $\ell = b_1(\Gamma)$. In this range, the algebraic differential form that is integrated in the parametric representation of the Feynman integral only has singularities along the graph hypersurface X_Γ, while the polynomial $P_\Gamma(t, p)$ only appears in the numerator. This range covers, in particular, the special case of the *log divergent* graphs. These are the graphs with $n = D\ell/2$, for which the integral (3.22) reduces to the simpler form

$$\mathcal{I}(\Gamma, p_1, \ldots, p_N) = \int_{\sigma_n} \frac{\omega_n}{\Psi_\Gamma(t)^{D/2}}. \tag{3.24}$$

We will not discuss the *unstable range* of small D, though it is often of significant physical interest, because of additional difficulties in treating the hypersurfaces defined by the second graph polynomial $P_\Gamma(t, p)$ caused by the additional dependence on the external momenta. We only make some general considerations about this case in §3.3 below.

In terms of graph hypersurfaces, the Feynman integral with the condition (3.23) can be regarded (modulo the problem of divergences) as a period obtained by integrating an algebraic differential form defined on the hypersurface complement $\mathbb{P}^{n-1} \smallsetminus X_\Gamma$ over a domain given by the topological simplex σ_n, whose boundary $\partial\sigma_n$ is contained in the normal crossings divisor Σ_n given by the algebraic simplex (the union of the coordinate hyperplanes). Thus, the motivic nature of this period should be detected by the mixed motive

$$\mathfrak{m}(\mathbb{P}^{n-1} \smallsetminus X_\Gamma, \Sigma_n \smallsetminus (\Sigma_n \cap X_\Gamma)), \tag{3.25}$$

whose realization is the relative cohomology

$$H^{n-1}(\mathbb{P}^{n-1} \smallsetminus X_\Gamma, \Sigma_n \smallsetminus (\Sigma_n \cap X_\Gamma)). \tag{3.26}$$

A possible strategy to showing that the Feynman integrals evaluate to numbers that are periods of mixed Tate motives would be to show that the motive (3.25) is mixed Tate. This is one of the main themes that we are going to develop in the rest of the book.

It is important to stress the fact that identifying the motivic nature of the relative cohomology above would only suffice up to the important issue of divergences. Divergences occur where the graph hypersurface X_Γ meets the domain of integration σ_n. Notice that the intersections $X_\Gamma \cap \sigma_n$ can only occur on the boundary $\partial \sigma_n$. In fact, in the interior of σ_n all the coordinates t_e are strictly positive real numbers, and the graph polynomial is then also strictly positive, $\Psi_\Gamma(t) > 0$, as can easily be seen by the expression (3.11). Thus, we have $X_\Gamma \cap \sigma_n \subset X_\Gamma \cap \Sigma_n$. These intersections are usually nonempty, so that some method of regularization needs to be introduced to remove the corresponding divergences in the integral, before one can identify the integration

$$\int_{\sigma_n} \frac{P_\Gamma(t,p)^{-n+D\ell/2} \omega_n}{\Psi_\Gamma(t)^{-n+D(\ell+1)/2}} \tag{3.27}$$

with a period of the motive (3.25). The regularization method proposed in [Bloch, Esnault, Kreimer (2006)] consists of performing a number of blow-ups of (strata of) the locus $X_\Gamma \cap \Sigma_n$, with the result of separating the domain of integration and the hypersurface and therefore regularizing the integral by replacing the original integration with one performed in the blown up variety. This regularization procedure adds a further complication to the problem of identifying the motivic nature of (3.25). Namely, one also needs to know that the motive remains mixed Tate after the blow-ups. We will return to this issue in more detail in §3.14 below, after we discuss the approach of [Aluffi and Marcolli (2009a)].

We now discuss briefly the difference between working with the affine or the projective hypersurfaces. While it is natural to work projectively when one has equations defined by homogeneous polynomials, there are various reasons why the affine context appears to be more natural in this case. In particular, we argue as in [Aluffi and Marcolli (2008b)] that the multiplicative property of the Feynman rules is more naturally reflected by a corresponding multiplicative property of the affine hypersurface complements, which in the projective case is replaced by the more complicated *join* operation.

As we saw in §1.3, the Feynman rules are multiplicative over disjoint unions of graphs, namely

$$U(\Gamma, p) = U(\Gamma_1, p_1) \cdots U(\Gamma_k, p_k), \tag{3.28}$$

for $\Gamma = \Gamma_1 \cup \cdots \cup \Gamma_k$ a disjoint union of graphs with $N_j = \#E_{ext}(\Gamma_j)$ external edges, with external momenta

$$p = (p_1, \ldots, p_k), \quad \text{with} \quad p_j = (p_{j,1}, \ldots, p_{j,N_j}).$$

The other "multiplicative" property that the Feynman rules satisfy is the one we discussed in the first chapter, which allows one to determine the

Feynman integrals of non-1PI graphs using the integrals for 1PI graphs. In fact, upon representing an arbitrary (finite) graph as a tree T with vertices $v \in V(T)$ replaced by 1PI graphs Γ_v with a number of external edges $\#E_{ext}(\Gamma_v) = \text{val}(v)$ equal to the valence of the vertex, one obtains the expression for the Feynman integral of the resulting graph Γ as

$$V(\Gamma, p) = \prod_{v \in V(T), e \in E_{int}(T), v \in \partial(e)} V(\Gamma_v, p_v) q_e(p_v)^{-1} \delta((p_v)_e - (p_{v'})_e), \quad (3.29)$$

where, as in (1.35) we write

$$V(\Gamma, p_1, \ldots, p_N) = \varepsilon(p_1, \ldots, p_N) U(\Gamma, p_1, \ldots, p_N),$$

with $\varepsilon(p_1, \ldots, p_N) = \prod_{e \in E_{ext}(\Gamma)} q_e(p_e)^{-1}$.

In the particular case of a massive theory $m \neq 0$ where one sets all the external momenta equal to zero, this becomes an actual product

$$U(\Gamma, p)|_{p=0} = U(e, p_e)|_{p_e=0}^{\#E_{int}(T)} \prod_{v \in V(T)} U(\Gamma_v, p_v)|_{p_v=0}, \quad (3.30)$$

where in this case the edge propagator $U(e, p_e)|_{p_e=0} = m^{-2}$ is just a constant depending on the mass parameter $m \neq 0$.

One can then, following [Aluffi and Marcolli (2008b)], define an *abstract Feynman rule* in the following way.

Definition 3.2.1. An abstract Feynman rule is an assignment of an element $U(\Gamma)$ in a commutative ring $\mathcal{R}$ for each finite graph Γ, with the property that, for a disjoint union $\Gamma = \Gamma_1 \cup \cdots \cup \Gamma_k$ this satisfies the multiplicative property

$$U(\Gamma) = U(\Gamma_1) \cdots U(\Gamma_k) \quad (3.31)$$

and for a non-1PI graph obtained by inserting 1PI graphs Γ_v at the vertices of a tree T it satisfies

$$U(\Gamma) = U(L)^{\#E_{int}(T)} \prod_{v \in V(T)} U(\Gamma_v). \quad (3.32)$$

A first difference one encounters between the projective graph hypersurfaces X_Γ and the affine $\hat{X}_\Gamma$, which is directly relevant to the quantum field theory setting, is the fact that the affine hypersurface complements behave multiplicatively as abstract Feynman rules are expected to do, while the projective hypersurface complements do not. More precisely, we have the following result from [Aluffi and Marcolli (2008b)].

Lemma 3.2.2. *Let $\Gamma = \Gamma_1 \cup \cdots \cup \Gamma_k$ be a disjoint union of finite graphs. Then the affine hypersurface complements satisfy*

$$\mathbb{A}^n \smallsetminus \hat{X}_\Gamma = (\mathbb{A}^{n_1} \smallsetminus \hat{X}_{\Gamma_1}) \times \cdots \times (\mathbb{A}^{n_k} \smallsetminus \hat{X}_{\Gamma_k}), \quad (3.33)$$

where $n_i = \#E_{int}(\Gamma_i)$.

Proof. Using the definition of the graph polynomials in terms of the determinant $\Psi_\Gamma(t) = \det \mathcal{M}_\Gamma(t)$ of the Kirchhoff matrix of the graph, one sees easily that if $\Gamma = \Gamma_1 \cup \cdots \cup \Gamma_k$ is a disjoint union, then one has

$$\Psi_\Gamma(t) = \prod_{i=1}^k \Psi_{\Gamma_1}(t_{i,1}, \ldots, t_{i,n_i}),$$

where $t = (t_e) = (t_{i,j})_{i=1,\ldots,k, j=1,\ldots,n_i}$ are the edge variables of the internal edges of Γ. This means that the affine hypersurface $\hat{X}_\Gamma$ is a union

$$\hat{X}_\Gamma = \bigcup_{i=1}^k (\mathbb{A}^{n_1} \times \cdots \times \mathbb{A}^{n_{i-1}} \times \hat{X}_{\Gamma_i} \times \mathbb{A}^{n_{i+1}} \times \cdots \times \mathbb{A}^{n_k}),$$

which implies that the hypersurface complement is then given by the product (3.33). $\qquad\square$

The situation with the projective hypersurfaces is more complicated. Instead of getting directly a product of the complements one gets a torus bundle. For instance, suppose given a disjoint union of two graphs $\Gamma = \Gamma_1 \cup \Gamma_2$ and assume, moreover, that neither graph is a forest (the assumption is not necessary in the affine case). Then the projective hypersurface complement $\mathbb{P}^{n_1+n_2-1} \smallsetminus X_\Gamma$ is a $\mathbb{G}_m$-bundle over the product $(\mathbb{P}^{n_1-1} \smallsetminus X_{\Gamma_1}) \times (\mathbb{P}^{n_2-1} \smallsetminus X_{\Gamma_2})$. To see this, observe that in this case the projective hypersurface X_Γ is a union of cones $C^{n_2}(X_{\Gamma_1})$ and $C^{n_1}(X_{\Gamma_2})$ in $\mathbb{P}^{n_1+n_2-1}$, respectively over X_{Γ_1} and X_{Γ_2} with vertices $\mathbb{P}^{n_2-1}$ and $\mathbb{P}^{n_1-1}$, respectively. The hypothesis that neither graph is a forest ensures that there is a regular map

$$\mathbb{P}^{n_1+n_2-1} \smallsetminus X_\Gamma \to (\mathbb{P}^{n_1-1} \smallsetminus X_{\Gamma_1}) \times (\mathbb{P}^{n_2-1} \smallsetminus X_{\Gamma_2}) \qquad (3.34)$$

given by

$$(t_{1,1} : \cdots : t_{1,n_1} : t_{2,1} : \cdots : t_{2,n_2}) \mapsto ((t_{1,1} : \cdots : t_{1,n_1}), (t_{2,1} : \cdots : t_{2,n_2})), \qquad (3.35)$$

while if either Γ_i happened to be a forest, the corresponding X_{Γ_i} would be empty, and the map (3.35) would not be defined everywhere. Under the assumption that neither Γ_i is a forest, the map (3.34) is surjective and the fiber over a point $((t_{1,1} : \cdots : t_{1,n_1}), (t_{2,1} : \cdots : t_{2,n_2}))$ is a copy of the multiplicative group $\mathbb{G}_m$ given by all the points of the form

$$(ut_{1,1} : \cdots : ut_{1,n_1} : vt_{2,1} : \cdots : vt_{2,n_2}), \quad \text{with} \quad (u : v) \in \mathbb{P}^1, \ uv \neq 0.$$

Equivalently, notice that if Γ is not a forest, then there is a regular map

$$\mathbb{A}^n \smallsetminus \hat{X}_\Gamma \to \mathbb{P}^{n-1} \smallsetminus X_\Gamma,$$

and (3.34) is then induced by the isomorphism in (3.33). This is no longer the case if Γ is a forest.

3.3 Landau varieties

For simplicity we restrict here everywhere to the case where the differential form in the parametric Feynman integral has singularities only along the graph hypersurface $\hat{X}_\Gamma$ defined by the vanishing of the graph polynomial $\Psi_\Gamma(t)$. This is certainly the case within what we referred to as the "stable range", *i.e.* for D (considered as a variable) sufficiently large so that $n \leq D\ell/2$. This, however, is typically not the range of physical interest for theories where $D = 4$ and the number of loops and internal edges violates the above inequality. Away from this stable range, the second graph polynomial $P_\Gamma(t, p)$ also appears in the denominator, and the differential form of the parametric Feynman integral is then defined on the complement of the hypersurface (family of hypersurfaces)

$$\hat{Y}_\Gamma(p) = \{t \in \mathbb{A}^n \mid P_\Gamma(t, p) = 0\} \tag{3.36}$$

or in projective space, since $P_\Gamma(\cdot, p)$ is homogeneous of degree $b_1(\Gamma) + 1$, on the complement of

$$Y_\Gamma(p) = \{t \in \mathbb{P}^{n-1} \mid P_\Gamma(t, p) = 0\}, \tag{3.37}$$

or else on the complement of the union of hypersurfaces $\hat{Y}_\Gamma(p) \cup \hat{X}_\Gamma$, again depending on the values of n, ℓ and D.

We only discuss here the case where $-n + D\ell/2 \geq 0$, though some of the same arguments extend to the other case as in [Marcolli (2008)]. The hypersurfaces $\hat{Y}_\Gamma(p)$ defined by the vanishing of $P_\Gamma(p, t)$, for a fixed value of the external momenta p, are sometimes referred to as the Landau varieties. The general type of arguments described above using the graph hypersurfaces $\hat{X}_\Gamma$ should then be adapted to the case of the Landau varieties. In such cases, however, the situation is more complicated because of the explicit dependence on the additional parameters p. In particular, some of the divergences of the integral, coming from the intersection of the hypersurface $Y_\Gamma(p)$ with the domain of integration σ_n, can occur in the interior of the domain, not only on its boundary as in the case of X_Γ; see [Bjorken and Drell (1965)], §18.

For simplicity, it is also convenient to make the following assumption on the polynomials $P_\Gamma(t, p)$ and Ψ_Γ, so that we avoid cancellations of common factors.

Definition 3.3.1. A 1PI graph Γ satisfies the *generic condition* on the external momenta if, for p in a dense open set in the space of external momenta, the polynomials $P_\Gamma(t, p)$ and $\Psi_\Gamma(t)$ have no common factor.

Recall that $P_\Gamma(t, p)$ is defined in terms of external momenta and cut-sets through the functions P_v and s_C as in Proposition 3.1.8. In terms of spanning trees, one has

$$P_\Gamma(t, p) = \sum_T \sum_{e' \in T} s_{T,e'} \, t_{e'} \prod_{e \in T^c} t_e, \qquad (3.38)$$

where $s_{T,e'} = s_C$ for the cut-set $C = T^c \cup \{e'\}$.

The parameterizing space of the external momenta is the hyperplane in the affine space $\mathbb{A}^{D \cdot \#E_{ext}(\Gamma)}$ obtained by imposing the conservation law

$$\sum_{e \in E_{ext}(\Gamma)} p_e = 0. \qquad (3.39)$$

Thus, the simplest possible configuration of external momenta is the one where one puts all the external momenta to zero, except for a pair $p_{e_1} = p = -p_{e_2}$ associated to a choice of a pair of external edges $\{e_1, e_2\} \subset E_{ext}(\Gamma)$. Let v_i be the unique vertex attached to the external edge e_i of the chosen pair. We then have, in this case, $P_{v_1} = p = -P_{v_2}$. Upon writing the polynomial $P_\Gamma(t, p)$ in the form (3.38), we obtain in this case

$$P_\Gamma(p, t) = p^2 \sum_T (\sum_{e' \in T_{v_1, v_2}} t_{e'}) \prod_{e \notin T} t_e, \qquad (3.40)$$

where $T_{v_1, v_2} \subset T$ is the unique path in T without backtrackings connecting the vertices v_1 and v_2. We use as in (3.18)

$$s_C = \left(\sum_{v \in V(\Gamma_1)} P_v \right)^2 = \left(\sum_{v \in V(\Gamma_2)} P_v \right)^2$$

to get $s_C = p^2$ for all the nonzero terms in this (3.40). These are all the terms that correspond to cut sets C such that the vertices v_1 and v_2 belong to different components. These cut sets consist of the complement of a spanning tree T and an edge of T_{v_1, v_2}. Consider for simplicity the case where the polynomial Ψ_Γ is irreducible. If we denote the linear functions of (3.40) by

$$L_T(t) = p^2 \sum_{e \in T_{v_1, v_2}} t_e, \qquad (3.41)$$

we see that, if the polynomial $\Psi_\Gamma(t)$ divides (3.40), one would have

$$P_\Gamma(p, t) = \Psi_\Gamma(t) \cdot L(t),$$

for a degree one polynomial $L(t)$, and this would give

$$\sum_T (L_T(t) - L(t)) \prod_{e \notin T} t_e \equiv 0,$$

for all t. One then sees, for example, that the 1PI condition on the graph Γ is necessary in order to have the condition of Definition 3.3.1. In fact, for a graph that is not 1PI, one may be able to find vertices and momenta as above such that the degree one polynomials $L_T(t)$ are all equal to the same $L(t)$. Generally, the validity of the condition of Definition 3.3.1 can be checked algorithmically to be satisfied for large classes of 1PI graphs, though a complete combinatorial characterization of the graphs satisfying these properties does not appear to be known.

3.4   Integrals in affine and projective spaces

As we have seen above, the homogeneous graph polynomial Ψ_Γ defines either a projective hypersurface $X_\Gamma \subset \mathbb{P}^{n-1}$, with $n = \#E_{int}(\Gamma)$, or an affine hypersurface $\hat{X}_\Gamma \subset \mathbb{A}^n$, the affine cone over X_Γ. Thus, assuming we are in the stable range of sufficiently high D where $-n + D\ell/2 \geq 0$, for $\ell = b_1(\Gamma)$ the number of loops, one can regard the Feynman integral $U(\Gamma, p)$ as computed over the affine hypersurface complement $\mathbb{A}^n \smallsetminus \hat{X}_\Gamma$, but one can also reformulate it in terms of the projective hypersurface complement.

In order then to reformulate in projective space $\mathbb{P}^{n-1}$ integrals originally defined in affine space $\mathbb{A}^n$, we first recall some basic facts about differential forms on affine and projective spaces and their relation, following mostly [Dimca (1992)], see also [Gelfand, Gindikin, and Graev (1980)].

The projective analog of the volume form

$$\omega_n = dt_1 \wedge \cdots \wedge dt_n$$

is given by the form

$$\Omega = \sum_{i=1}^{n} (-1)^{i+1} t_i \, dt_1 \wedge \cdots \wedge \widehat{dt_i} \wedge \cdots \wedge dt_n. \tag{3.42}$$

The relation between the volume form $dt_1 \wedge \cdots \wedge dt_n$ and the homogeneous form Ω of degree n of (3.42) is given by

$$\Omega = \Delta(\omega_n), \tag{3.43}$$

where $\Delta : \Omega^k \to \Omega^{k-1}$ is the operator of contraction with the Euler vector field

$$E = \sum_i t_i \frac{\partial}{\partial t_i}, \tag{3.44}$$

$$\Delta(\omega)(v_1, \ldots, v_{k-1}) = \omega(E, v_1, \ldots, v_{k-1}). \tag{3.45}$$

Let $\mathcal{R} = \mathbb{C}[t_1, \ldots, t_n]$ be the ring of polynomials of $\mathbb{A}^n$. Let $\mathcal{R}_m$ denote the subset of homogeneous polynomials of degree m. Similarly, let Ω^k denote the $\mathcal{R}$-module of k-forms on $\mathbb{A}^n$ and let Ω^k_m denote the subset of k-forms that are homogeneous of degree m.

Let f be a homogeneous polynomial function $f : \mathbb{A}^n \to \mathbb{A}$ of degree $\deg(f)$. Let $\pi : \mathbb{A}^n \smallsetminus \{0\} \to \mathbb{P}^{n-1}$ be the standard projection $t = (t_1, \ldots, t_n) \mapsto \pi(t) = (t_1 : \cdots : t_n)$. We denote by $\hat{X}_f = \{t \in \mathbb{A}^n \mid f(t) = 0\}$ the affine hypersurface and by $X_f = \{\pi(t) \in \mathbb{P}^{n-1} \mid f(t) = 0\}$ the projective hypersurface, and by $\hat{\mathcal{D}}(f) = \mathbb{A}^n \smallsetminus \hat{X}_f$ and $\mathcal{D}(f) = \mathbb{P}^{n-1} \smallsetminus X_f$ the hypersurface complements. With the notation introduced above, we can always write a form $\omega \in \Omega^k(\mathcal{D}(f))$ as

$$\omega = \frac{\eta}{f^m}, \quad \text{with} \quad \eta \in \Omega^k_{m \deg(f)}. \tag{3.46}$$

We then have the following characterization of the pullback along $\pi :$ $\hat{\mathcal{D}}(f) \to \mathcal{D}(f)$ of forms on $\mathcal{D}(f)$.

Proposition 3.4.1. (see [Dimca (1992)], p.180 and [Dolgachev (1982)]) *Given $\omega \in \Omega^k(\mathcal{D}(f))$, the pullback $\pi^*(\omega) \in \Omega^k(\hat{\mathcal{D}}(f))$ is characterized by the properties of being invariant under the $\mathbb{G}_m$ action on $\mathbb{A}^n \smallsetminus \{0\}$ and of satisfying $\Delta(\pi^*(\omega)) = 0$, where Δ is the contraction (3.45) with the Euler vector field E of (3.44).*

Thus, since the sequence

$$0 \to \Omega^n \xrightarrow{\Delta} \Omega^{n-1} \xrightarrow{\Delta} \cdots \xrightarrow{\Delta} \Omega^1 \xrightarrow{\Delta} \Omega^0 \to 0$$

is exact at all but the last term, one can write

$$\pi^*(\omega) = \frac{\Delta(\eta)}{f^m}, \quad \text{with} \quad \eta \in \Omega^k_{m \deg(f)}. \tag{3.47}$$

In particular, any $(n-1)$-form on $\mathcal{D}(f) \subset \mathbb{P}^{n-1}$ can be written as

$$\frac{h\Omega}{f^m}, \quad \text{with} \quad h \in \mathcal{R}_{m \deg(f) - n} \tag{3.48}$$

and with $\Omega = \Delta(dt_1 \wedge \cdots \wedge dt_n)$ the $(n-1)$-form (3.42), homogeneous of degree n.

We then obtain the following result formulating integrals in affine spaces in terms of integrals of pullbacks of forms from projective spaces.

Proposition 3.4.2. *Let $\omega \in \Omega^k_{m \deg(f)}$ be a closed k-form, which is homogeneous of degree $m \deg(f)$, and consider the form ω/f^m on $\mathbb{A}^n$. Let*

$\sigma \subset \mathbb{A}^n \setminus \{0\}$ *be a* k-*dimensional domain with boundary* $\partial\sigma \neq \emptyset$. *Then the integration of* ω/f^m *over* σ *satisfies*

$$m \deg(f) \int_\sigma \frac{\omega}{f^m} = \int_{\partial\sigma} \frac{\Delta(\omega)}{f^m} + \int_\sigma df \wedge \frac{\Delta(\omega)}{f^{m+1}}. \tag{3.49}$$

Proof. Recall that we have ([Dimca (1992)], [Dolgachev (1982)])

$$d\left(\frac{\Delta(\omega)}{f^m}\right) = -\frac{\Delta(d_f\omega)}{f^{m+1}}, \tag{3.50}$$

where, for a form ω that is homogeneous of degree $m \deg(f)$,

$$d_f\omega = f \, d\omega - m \, df \wedge \omega. \tag{3.51}$$

Thus, we have

$$d\left(\frac{\Delta(\omega)}{f^m}\right) = -\frac{\Delta(d\omega)}{f^m} + m\frac{\Delta(df \wedge \omega)}{f^{m+1}}. \tag{3.52}$$

Since the form ω is closed, $d\omega = 0$, and we have

$$\Delta(df \wedge \omega) = \deg(f) \, f \, \omega - df \wedge \Delta(\omega), \tag{3.53}$$

we obtain from the above

$$d\left(\frac{\Delta(\omega)}{f^m}\right) = m \deg(f)\frac{\omega}{f^m} - \frac{df \wedge \Delta(\omega)}{f^{m+1}}. \tag{3.54}$$

By Stokes' theorem we have

$$\int_{\partial\sigma} \frac{\Delta(\omega)}{f^m} = \int_\sigma d\left(\frac{\Delta(\omega)}{f^m}\right).$$

Using (3.54) this gives

$$\int_{\partial\sigma} \frac{\Delta(\omega)}{f^m} = m \deg(f) \int_\sigma \frac{\omega}{f^m} - \int_\sigma \frac{df \wedge \Delta(\omega)}{f^{m+1}}.$$

which gives (3.49). $\qquad\square$

In the range $-n + D(\ell + 1)/2 \geq 0$ we consider the Feynman integral $U(\Gamma, p)$ as defined on the affine hypersurface complement $\mathbb{A}^n \setminus \hat{X}_\Gamma$ and we use the result above to reformulate the parametric Feynman integrals in terms of integrals of forms that are pullbacks to $\mathbb{A}^n \setminus \{0\}$ of forms on a hypersurface complement in $\mathbb{P}^{n-1}$. For simplicity, we remove here the divergent Γ-factor from the parametric Feynman integral and we concentrate on the residue given by the integration over the simplex σ as in (3.55) below.

Proposition 3.4.3. *Under the generic condition on the external momenta, and assuming that $-n + D\ell/2 \geq 0$, the parametric Feynman integral*

$$U(\Gamma, p) = \int_\sigma \frac{V_\Gamma^{-n+D\ell/2} \omega_n}{\Psi_\Gamma^{D/2}}, \tag{3.55}$$

with $V_\Gamma(t, p) = P_\Gamma(t, p)/\Psi_\Gamma(t, p)$, can be computed as

$$U(\Gamma, p) = \frac{1}{C(n, D, \ell)} \left(\int_{\partial\sigma} \pi^*(\eta) + \int_\sigma df \wedge \frac{\pi^*(\eta)}{f} \right), \tag{3.56}$$

where $\pi : \mathbb{A}^n \smallsetminus \{0\} \to \mathbb{P}^{n-1}$ is the projection and η is the form on the hypersurface complement $\mathcal{D}(f)$ in $\mathbb{P}^{n-1}$ with

$$\pi^*(\eta) = \frac{\Delta(\omega)}{f^m}, \tag{3.57}$$

on $\mathbb{A}^n$, where $f = \Psi_\Gamma$ and $m = -n + D(\ell+1)/2$, and $\omega = P_\Gamma(t, p)^{-n+D\ell/2} \omega_n$ with $\omega_n = dt_1 \wedge \cdots \wedge dt_n$ the volume form on $\mathbb{A}^n$. The coefficient $C(n, D, \ell)$ in (3.56) is given by

$$C(n, D, \ell) = (-n + D(\ell+1)/2)\ell. \tag{3.58}$$

Proof. Consider on $\mathbb{A}^n$ the form given by $\Delta(\omega)/f^m$, with f, m, and ω as above. We assume the condition of Definition 3.3.1, *i.e.* for a generic choice of the external momenta the polynomials P_Γ and Ψ_Γ have no common factor. First notice that, since the polynomial Ψ_Γ is homogeneous of degree ℓ and P_Γ is homogeneous of degree $\ell+1$, the form $\Delta(\omega)/f^m$ is $\mathbb{G}_m$-invariant on $\mathbb{A}^n \smallsetminus \{0\}$. Moreover, since it is of the form $\alpha = \Delta(\omega)/f^m$, it also satisfies $\Delta(\alpha) = 0$, hence it is the pullback of a form η on $\mathcal{D}(f) \subset \mathbb{P}^{n-1}$. Also notice that the domain of integration $\sigma \subset \mathbb{A}^n$, given by the simplex $\sigma = \sigma_n = \{\sum_i t_i = 1, t_i \geq 0\}$, is contained in a fundamental domain of the action of the multiplicative group $\mathbb{C}^*$ on $\mathbb{C}^n \smallsetminus \{0\}$.

Applying the result of Proposition 3.4.2 above, we obtain

$$\int_\sigma \frac{P_\Gamma(t, p)^{-n+D\ell/2} dt_1 \wedge \cdots \wedge dt_n}{\Psi_\Gamma^{-n+D(\ell+1)/2}} = \int_\sigma \frac{\omega}{f^m}$$

$$= \frac{1}{m \deg(f)} \left(\int_{\partial\sigma} \frac{\Delta(\omega)}{f^m} + \int_\sigma df \wedge \frac{\Delta(\omega)}{f^{m+1}} \right)$$

$$= C(n, D, \ell)^{-1} \left(\int_{\partial\sigma} \frac{P_\Gamma(t, p)^a \Delta(\omega_n)}{\Psi_\Gamma^m} + \int_\sigma df \wedge \frac{P_\Gamma(t, p)^a \Delta(\omega_n)}{\Psi_\Gamma^{m+1}} \right),$$

with $a = -n + D\ell/2$ and $m = -n + D(\ell+1)/2$. The coefficient $C(n, D, \ell)$ is given by $C(n, D, \ell) = m \deg(f)$, with m and f as above, hence it is given by (3.58). $\qquad\square$

3.5 Non-isolated singularities

The graph hypersurfaces $X_\Gamma \subset \mathbb{P}^{n-1}$ defined by the vanishing of the polynomial $\Psi_\Gamma(t) = \det(M_\Gamma(t))$ usually have non-isolated singularities. This can easily be seen by the following observation.

Lemma 3.5.1. *Let Γ be a graph with $\deg \Psi_\Gamma > 2$. The singular locus of X_Γ is given by the intersection of cones over the hypersurfaces X_{Γ_e}, for $e \in E(\Gamma)$, where Γ_e is the graph obtained by removing the edge e of Γ. The cones $C(X_{\Gamma_e})$ do not intersect transversely.*

Proof. First observe that, since X_Γ is defined by a homogeneous equation $\Psi_\Gamma(t) = 0$, with Ψ_Γ a polynomial of degree m, the Euler formula $m\Psi_\Gamma(t) = \sum_e t_e \frac{\partial}{\partial t_e} \Psi_\Gamma(t)$ implies that $\cap_e Z(\partial_e \Psi_\Gamma) \subset X_\Gamma$, where $Z(\partial_e \Psi_\Gamma)$ is the zero locus of the t_e-derivative. Thus, the singular locus of X_Γ is just given by the equations $\partial_e \Psi_\Gamma = 0$. The variables t_e appear in the polynomial $\Psi_\Gamma(t)$ only with degree zero or one, hence the polynomial $\partial_e \Psi_\Gamma$ consists of only those monomials of Ψ_Γ that contain the variable t_e, where one sets $t_e = 1$. The resulting polynomial is therefore of the form Ψ_{Γ_e}, where Γ_e is the graph obtained from Γ by removing the edge e. In fact, one can see in terms of spanning trees that, if T is a spanning tree containing the edge e, then $T \smallsetminus e$ is no longer a spanning tree of Γ_e, so the corresponding terms disappear in passing from Ψ_Γ to Ψ_{Γ_e}, while if T is a spanning tree of Γ which does not contain e, then T is still a spanning tree of Γ_e and the corresponding monomial m_T of Ψ_{Γ_e} is the same as the monomial m_T in Ψ_Γ without the variable t_e. Thus, the zero locus $Z(\Psi_{\Gamma_e}) \subset \mathbb{P}^{n-1}$ is a cone $C(X_{\Gamma_e})$ over the graph hypersurface $X_{\Gamma_e} \subset \mathbb{P}^{n-2}$ with vertex at the coordinate point $v_e = (0, \ldots, 0, 1, 0, \ldots 0)$ with $t_e = 1$. To see that these cones do not intersect transversely, notice that, in the case where $\deg \Psi_\Gamma > 2$, given any two $C(X_{\Gamma_e})$ and $C(X_{\Gamma_{e'}})$, the vertex of one cone is contained in the graph hypersurface spanning the other cone. $\qquad \square$

In fact, the hypersurfaces X_Γ tend to have singularity loci of low codimension. As we observe again later, this has important consequences from the motivic viewpoint. Moreover, it makes it especially useful to adopt methods from singularity theory to study these hypersurfaces. For example, we are going to see in the following sections how characteristic classes of singular varieties can be employed in this context. These provide a way of measuring how singular a hypersurface is, as we discuss more in detail in §3.9 below. Ongoing work [Bergbauer and Rej (2009)] gives an analysis of the singular locus of the graph hypersurfaces using a formula for the Kirchhoff polynomials Ψ_Γ under insertion of subgraphs at vertices.

3.6   Cremona transformation and dual graphs

A useful tool to study the algebraic geometry of the projective graph hypersurfaces X_Γ is the Cremona transformation, which for planar graphs relates the hypersurface of a graph with that of its dual. This was pointed out in [Bloch (2007)] and we illustrate the principle here in the particular completely explicit case of the "banana graphs" computed in [Aluffi and Marcolli (2008a)]. Even for graphs that are non-planar, one can still use the image of the graph hypersurface under the Cremona transformation to derive useful information on its motivic nature, as shown in the recent results of [Bloch (2008)]. We follow [Aluffi and Marcolli (2008a)].

Definition 3.6.1. The standard Cremona transformation of $\mathbb{P}^{n-1}$ is the map

$$\mathcal{C} : (t_1 : \cdots : t_n) \mapsto \left(\frac{1}{t_1} : \cdots : \frac{1}{t_n} \right). \tag{3.59}$$

Let $\mathcal{G}(\mathcal{C})$ denote the closure of the graph of $\mathcal{C}$. Then $\mathcal{G}(\mathcal{C})$ is a subvariety of $\mathbb{P}^{n-1} \times \mathbb{P}^{n-1}$ with projections

$$\mathcal{G}(\mathcal{C}) \tag{3.60}$$

$$\mathbb{P}^{n-1} \xleftarrow{\;\;\pi_1\;\;} \quad \xrightarrow{\;\;\pi_2\;\;} \mathbb{P}^{n-1}$$

$$\mathbb{P}^{n-1} \dashrightarrow{\;\;\;\mathcal{C}\;\;\;} \mathbb{P}^{n-1}$$

We introduce the following notation. Let $\mathcal{S}_n \subset \mathbb{P}^{n-1}$ be defined by the ideal

$$\mathcal{I}_{\mathcal{S}_n} = (t_1 \cdots t_{n-1}, t_1 \cdots t_{n-2} t_n, \ldots, t_1 t_3 \cdots t_n, t_2 \cdots t_n). \tag{3.61}$$

Also, let $\mathcal{L}$ be the hyperplane defined by the equation

$$\mathcal{L} = \{ (t_1 : \cdots : t_n) \in \mathbb{P}^{n-1} \,|\, t_1 + \cdots + t_n = 0 \}. \tag{3.62}$$

The Cremona transformation is a priori defined only away from the algebraic simplex of coordinate axes

$$\Sigma_n := \{ (t_1 : \cdots : t_n) \in \mathbb{P}^{n-1} \,|\, \prod_i t_i = 0 \} \subset \mathbb{P}^{n-1}, \tag{3.63}$$

though, as we now show, it is in fact well defined also on the general point of Σ_n, its locus of indeterminacies being only the singularity subscheme $\mathcal{S}_n$ of the algebraic simplex Σ_n.

Lemma 3.6.2. *Let $\mathcal{C}$, $\mathcal{G}(\mathcal{C})$, $\mathcal{S}_n$, and $\mathcal{L}$ be as above.*

(1) The indeterminacy locus of $\mathcal{C}$ is $\mathcal{S}_n$.
(2) $\pi_1 : \mathcal{G}(\mathcal{C}) \to \mathbb{P}^{n-1}$ is the blow-up along $\mathcal{S}_n$.
(3) $\mathcal{L}$ intersects every component of $\mathcal{S}_n$ transversely.
(4) Σ_n cuts out a divisor with simple normal crossings on $\mathcal{L}$.

Proof. The components of the map defining $\mathcal{C}$ given in (3.59) can be rewritten as

$$(t_1 : \cdots : t_n) \mapsto (t_2 \cdots t_n : t_1 t_3 \cdots t_n : \cdots : t_1 \cdots t_{n-1}). \qquad (3.64)$$

Using (3.64), the map $\pi_1 : \mathcal{G}(\mathcal{C}) \to \mathbb{P}^n$ may be identified with the blow-up of $\mathbb{P}^n$ along the subscheme $\mathcal{S}_n$ defined by the ideal $\mathcal{I}_{\mathcal{S}_n}$ of (3.61) generated by the partial derivatives of the equation of the algebraic simplex (*i.e.* the singular locus of Σ_n). The properties (3) and (4) then follow. $\qquad \square$

Properties (3) and (4) of the proposition above only play a role implicitly in the proof of Proposition 3.7.1 below, so we leave here the details to the reader.

An example where it is easier to visualize the effect of the Cremona transformation is the case where $n = 3$. In this case, one can view the algebraic simplex Σ_3 in the projective plane $\mathbb{P}^2$ as a union of three lines, each two intersecting in a point. These three points, the zero-dimensional strata of Σ_3, are blown up in $\mathcal{G}(\mathcal{C})$ and each replaced by a line. The proper transforms of the 1-dimensional components of Σ_3 are then blown down by the other map π_2 of (3.60), so that the resulting image is again a triangle of three lines in $\mathbb{P}^2$. The Cremona transformation (the horizontal rational map in the diagram (3.60)) is an isomorphism of the complements of the two triangles.

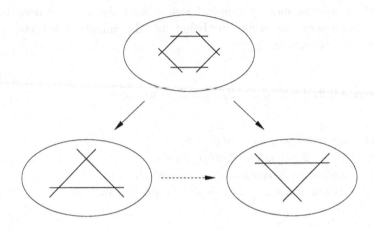

The *dual graph* $\Gamma^\vee$ of a planar graph Γ is defined by choosing an embedding $\iota : \Gamma \hookrightarrow S^2$ in the sphere and then taking as vertices $v \in V(\Gamma^\vee)$ a point in each connected component of $S^2 \smallsetminus \iota(\Gamma)$,

$$\#V(\Gamma^\vee) = b_0(S^2 \smallsetminus \iota(\Gamma)), \tag{3.65}$$

and edges $E(\Gamma^\vee)$ given by adding an edge between two vertices $v, w \in V(\Gamma^\vee)$ for each edge of Γ that is in the boundary between the regions of $S^2 \smallsetminus \iota(\Gamma)$ containing the points v and w. Thus, the dual graph has

$$\#E(\Gamma^\vee) = \#E(\Gamma). \tag{3.66}$$

Notice that it is somewhat misleading to talk about *the* dual graph $\Gamma^\vee$, since in fact this depends not only on Γ itself, but also on the choice of the embedding $\iota : \Gamma \hookrightarrow S^2$. One can give examples of different embeddings ι_1, ι_2 of the same graph Γ, for which the corresponding dual graphs $\Gamma_1^\vee$ and $\Gamma_2^\vee$ are topologically inequivalent, as the following figure shows.

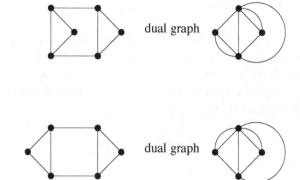

It was shown in [Bloch (2007)] that the Cremona transformation relates the graph hypersurfaces of Γ and its dual $\Gamma^\vee$ in the following way (see also [Aluffi and Marcolli (2008a)]).

Lemma 3.6.3. *Suppose given a planar graph Γ with $\#E(\Gamma) = n$, with dual graph $\Gamma^\vee$. Then the graph polynomials satisfy*

$$\Psi_\Gamma(t_1, \ldots, t_n) = (\prod_{e \in E(\Gamma)} t_e) \, \Psi_{\Gamma^\vee}(t_1^{-1}, \ldots, t_n^{-1}), \qquad (3.67)$$

hence the graph hypersurfaces are related by the Cremona transformation $\mathcal{C}$ of (3.59),

$$\mathcal{C}(X_\Gamma \cap (\mathbb{P}^{n-1} \smallsetminus \Sigma_n)) = X_{\Gamma^\vee} \cap (\mathbb{P}^{n-1} \smallsetminus \Sigma_n). \qquad (3.68)$$

Proof. This follows from the combinatorial identity

$$\Psi_\Gamma(t_1, \ldots, t_n) = \sum_{T \subset \Gamma} \prod_{e \notin E(T)} t_e$$
$$= (\prod_{e \in E(\Gamma)} t_e) \sum_{T \subset \Gamma} \prod_{e \in E(T)} t_e^{-1}$$
$$= (\prod_{e \in E(\Gamma)} t_e) \sum_{T' \subset \Gamma^\vee} \prod_{e \notin E(T')} t_e^{-1}$$
$$= (\prod_{e \in E(\Gamma)} t_e) \Psi_{\Gamma^\vee}(t_1^{-1}, \ldots, t_n^{-1}).$$

The third equality uses the fact that $\#E(\Gamma) = \#E(\Gamma^\vee)$ and $\#V(\Gamma^\vee) = b_0(S^2 \smallsetminus \Gamma)$, so that $\deg \Psi_\Gamma + \deg \Psi_{\Gamma^\vee} = \#E(\Gamma)$, and the fact that there is a bijection between complements of spanning tree T in Γ and spanning trees T' in $\Gamma^\vee$ obtained by shrinking the edges of T in Γ and taking the dual graph of the resulting connected graph. Written in the coordinates $(s_1 : \cdots : s_n)$ of the target $\mathbb{P}^{n-1}$ of the Cremona transformation, the identity (3.67) gives

$$\Psi_\Gamma(t_1, \ldots, t_n) = (\prod_{e \in E(\Gamma^\vee)} s_e^{-1}) \Psi_{\Gamma^\vee}(s_1, \ldots, s_n)$$

from which (3.68) follows. $\qquad\qquad\qquad\square$

3.7   Classes in the Grothendieck ring

An example of an infinite family of Feynman graphs for which the classes in the Grothendieck ring $K_0(\mathcal{V})$ can be computed explicitly was considered in [Aluffi and Marcolli (2008a)]. These are known as the *banana graphs*: for $n \geq 2$ the graph Γ_n in this family has two vertices of valence n and n parallel edges between them.

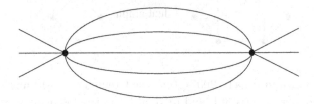

The banana graph Γ_n has graph polynomial

$$\Psi_\Gamma(t) = t_1 \cdots t_n \left(\frac{1}{t_1} + \cdots + \frac{1}{t_n} \right).$$

Thus, the parametric integral in this case is

$$\int_{\sigma_n} \frac{(t_1 \cdots t_n)^{(\frac{D}{2}-1)(n-1)-1} \omega_n}{\Psi_\Gamma(t)^{(\frac{D}{2}-1)n}},$$

up to an overall multiplicative factor proportional to the sum of external momenta at a vertex.

We gave in [Aluffi and Marcolli (2008a)] an explicit formula for the class in the Grothendieck ring of the graph hypersurfaces for the banana graphs. This is given by the following result.

Proposition 3.7.1. *The class in the Grothendieck ring of the hypersurface X_{Γ_n} of the n-th banana graph is explicitly given by the formula*

$$[X_{\Gamma_n}] = \frac{\mathbb{L}^n - 1}{\mathbb{L} - 1} - \frac{(\mathbb{L} - 1)^n - (-1)^n}{\mathbb{L}} - n\,(\mathbb{L} - 1)^{n-2}. \tag{3.69}$$

In particular, this formula shows that, in this example, the class $[X_{\Gamma_n}]$ manifestly lies in the mixed Tate part $\mathbb{Z}[\mathbb{L}]$ of the Grothendieck ring. We sketch the proof here and refer the reader to [Aluffi and Marcolli (2008a)] for more details.

Proof.   The formula (3.69) is proved using the method of Cremona transformation and dual graphs described above. In fact, one observes easily that, in the case of the banana graphs, the dual graph $\Gamma_n^\vee$ is simply a polygon with n sides, and therefore the associated graph hypersurface $X_{\Gamma_n^\vee} = \mathcal{L}$ is just a hyperplane in $\mathbb{P}^{n-1}$.

One then checks that the complement in this hyperplane of the algebraic simplex has class

$$[\mathcal{L} \smallsetminus (\mathcal{L} \cap \Sigma_n)] = [\mathcal{L}] - [\mathcal{L} \cap \Sigma_n] = \frac{\mathbb{T}^{n-1} - (-1)^{n-1}}{\mathbb{T} + 1}$$

where $\mathbb{T} = [\mathbb{G}_m] = [\mathbb{A}^1] - [\mathbb{A}^0]$ is the class of the multiplicative group. Moreover, one finds that $X_{\Gamma_n} \cap \Sigma_n = \mathcal{S}_n$, the scheme of singularities of Σ_n. The latter has class

$$[\mathcal{S}_n] = [\Sigma_n] - n\,\mathbb{T}^{n-2}.$$

This then gives

$$[X_{\Gamma_n}] = [X_{\Gamma_n} \cap \Sigma_n] + [X_{\Gamma_n} \smallsetminus (X_{\Gamma_n} \cap \Sigma_n)],$$

where one uses the Cremona transformation to identify

$$[X_{\Gamma_n}] = [\mathcal{S}_n] + [\mathcal{L} \smallsetminus (\mathcal{L} \cap \Sigma_n)].$$

$\square$

In particular, since the class in the Grothendieck ring is a universal Euler characteristic, this calculation also yields as a consequence the value for the topological Euler characteristic of the (complex) variety $X_{\Gamma_n}(\mathbb{C})$, which is of the form

$$\chi(X_{\Gamma_n}) = n + (-1)^n,$$

for $n \geq 3$, as one can see from the fact that the Euler characteristic is a ring homomorphism from $K_0(\mathcal{V})$ to $\mathbb{Z}$ and that in (3.69) $\mathbb{L} = [\mathbb{A}^1]$ has Euler characteristic $\chi(\mathbb{A}^1) = 1$. A different computation of the Euler characteristic $\chi(X_{\Gamma_n})$, based on the use of characteristic classes of singular varieties, is also given in [Aluffi and Marcolli (2008a)] along with the more complicated explicit computation of the Chern–Schwartz–MacPherson characteristic class of the hypersurfaces X_{Γ_n}.

More generally, the use of the Cremona transformation method can lead to interesting results even in the case where the graphs are not necessarily planar. In fact, one can still consider a *dual hypersurface* $X_\Gamma^\vee$ obtained as the image of X_Γ under the Cremona transformation and still have an isomorphism of X_Γ and $X_\Gamma^\vee$ away from the algebraic simplex Σ_n,

$$X_\Gamma \smallsetminus (X_\Gamma \cap \Sigma_n) \overset{\mathcal{C}}{\simeq} X_\Gamma^\vee \smallsetminus (X_\Gamma^\vee \cap \Sigma_n),$$

although in non-planar cases $X_\Gamma^\vee$ is no longer identified with the graph hypersurface of a dual graph $X_{\Gamma^\vee}$. By applying this method to the complete graph Γ_N on N vertices, Bloch proved in [Bloch (2008)] a very interesting result showing that, although the individual hypersurfaces X_Γ of Feynman graphs are not always mixed Tate motives, if one sums the classes over graphs with a fixed number of vertices one obtains a class that is always in the Tate part $\mathbb{Z}[\mathbb{L}]$ of the Grothendieck ring. More precisely, it is shown in [Bloch (2008)] that the class

$$S_N = \sum_{\#V(\Gamma)=N} [X_\Gamma] \frac{N!}{\#\mathrm{Aut}(\Gamma)},$$

where the sum is over all graphs with a fixed number of vertices, is always in $\mathbb{Z}[\mathbb{L}]$. This shows that there will be very interesting cancellations between the classes $[X_\Gamma]$ at least for a sufficiently large number of loops, where one knows by the general result of [Belkale and Brosnan (2003a)] that non-mixed-Tate contributions will eventually appear.

This result fits in well with the general fact that, in quantum field theory, individual Feynman graphs do not represent observable physical processes and only sums over graphs (usually with fixed external edges and external momenta) can be physically meaningful. Although in the more physical setting it would be natural to sum over graphs with a given number of loops and with given external momenta, rather than over a fixed number of vertices, the result of [Bloch (2008)] suggests that a more appropriate formulation of the conjecture on Feynman integrals and motives should perhaps be given directly in terms that involve the full expansion of perturbative quantum field theory, with sums over graphs, rather than in terms of individual graphs. This approach via families of graphs also fits in well with the treatment of gauge theories and Slavonov–Taylor identities within the context of the Connes–Kreimer approach to renormalization via Hopf algebras, as in [van Suijlekom (2007)], [van Suijlekom (2006)] and also [Kreimer and van Suijlekom (2009)] in the setting of Dyson–Schwinger equations.

3.8   Motivic Feynman rules

We have discussed above a definition of "abstract Feynman rules" based on the multiplicative properties (3.31) and (3.32) that express the expectation value of Feynman integrals associated to arbitrary Feynman graphs of a given scalar quantum field theory first in terms of connected graphs and then in terms of 1PI graphs.

This more abstract definition of Feynman rules allows one to make sense of algebro-geometric and motivic $U(\Gamma)$, with the same formal properties as the original Feynman integrals with respect to the combinatorics of graphs. As we discuss later on in the book, this will be closely related to defining Feynman rules in terms of characters of the Connes–Kreimer Hopf algebra of Feynman graphs and to the abstract form of the Birkhoff factorization that gives the renormalization procedure for Feynman integrals.

One can first observe, as in [Aluffi and Marcolli (2008b)], that the hypersurface complement $\mathbb{A}^n \smallsetminus \hat{X}_\Gamma$ where the parametric Feynman integral is computed behaves itself like a Feynman integral, in the sense that it satisfies the multiplicative properties (3.31) and (3.32). More precisely, we have seen in Lemma 3.2.2 above that for a disjoint union of graphs $\Gamma = \Gamma_1 \cup \Gamma_2$ the hypersurface complement splits multiplicatively as

$$\mathbb{A}^n \smallsetminus \hat{X}_\Gamma = (\mathbb{A}^{n_1} \smallsetminus \hat{X}_{\Gamma_1}) \times (\mathbb{A}^{n_2} \smallsetminus \hat{X}_{\Gamma_2}).$$

Lemma 3.2.2 shows that the hypersurface complement $\mathbb{A}^n \smallsetminus \hat{X}_\Gamma$ satisfies the multiplicative property (3.31), when we think of the Cartesian product $(\mathbb{A}^{n_1} \smallsetminus \hat{X}_{\Gamma_1}) \times (\mathbb{A}^{n_2} \smallsetminus \hat{X}_{\Gamma_2})$ as defining the product operation in a suitable commutative ring. We will specify the ring more precisely below as a refinement of the Grothendieck ring of varieties, namely the *ring of immersed conical varieties* introduced in [Aluffi and Marcolli (2008b)]. One can similarly see that the second property of Feynman rules, the multiplicative property (3.32) reducing connected graphs to 1PI graphs and inverse propagators, is also satisfied by the hypersurface complements.

Lemma 3.8.1. *For a connected graph Γ that is obtained by inserting 1PI graphs Γ_v at the vertices of a finite tree T, the hypersurface complement satisfies*

$$\mathbb{A}^n \smallsetminus \hat{X}_\Gamma = \mathbb{A}^{\#E(T)} \times \prod_{v \in V(T)} (\mathbb{A}^{\#E_{int}(\Gamma_v)} \smallsetminus \hat{X}_{\Gamma_v}). \tag{3.70}$$

Proof. If Γ is a forest with $n = \#E(\Gamma)$, then $\mathbb{A}^n \smallsetminus \hat{X}_\Gamma = \mathbb{A}^n$. Also notice that if Γ_1 and Γ_2 are graphs joined at a single vertex, then the same argument given in Lemma 3.2.2 for the disjoint union still gives the desired result. Thus, in the case of a connected graphs Γ obtained by inserting 1PI graphs at vertices of a tree, these two observations combine to give

$$(\mathbb{A}^{n_1} \smallsetminus \hat{X}_{\Gamma_{v_1}}) \times \cdots \times (\mathbb{A}^{n_{\#V(T)}} \smallsetminus \hat{X}_{\Gamma_{v_{\#V(T)}}}) \times \mathbb{A}^{\#E(T)}. \qquad \square$$

In the above, one can consider the hypersurface complements $\mathbb{A}^n \smallsetminus \hat{X}_\Gamma$ modulo certain equivalence relations. For example, if these varieties are considered up to isomorphism, and one imposes the inclusion–exclusion relation $[X] = [X \smallsetminus Y] + [Y]$ for closed subvarieties $Y \subset X$, one can define a Feynman rule with values in the Grothendieck ring of varieties

$$\mathbb{U}(\Gamma) := [\mathbb{A}^n \smallsetminus \hat{X}_\Gamma] = [\mathbb{A}^n] - [\hat{X}_\Gamma] \in K_0(\mathcal{V}). \qquad (3.71)$$

This satisfies (3.31) and (3.32) as an immediate consequence of Lemma 3.2.2 and Lemma 3.8.1.

The relation between the classes in the Grothendieck ring of the affine and projective hypersurface complements is given by the formula

$$[\mathbb{A}^n \smallsetminus \hat{X}_\Gamma] = (\mathbb{L} - 1)[\mathbb{P}^{n-1} \smallsetminus X_\Gamma],$$

which expresses the fact that $\hat{X}_\Gamma$ is the affine cone over X_Γ.

One can also impose a weaker equivalence relation, by identifying varieties not up to isomorphism, but only up to linear changes of coordinates in an ambient affine space. This leads to the following refinement of the Grothendieck ring of varieties ([Aluffi and Marcolli (2008b)]).

Definition 3.8.2. The *ring of immersed conical varieties* $\mathcal{F}_{\mathbb{K}}$ is generated by classes $[V]$ of equivalence under linear coordinate changes of varieties $V \subset \mathbb{A}^N$ defined by homogeneous ideals (hence the name "conical") immersed in some arbitrarily large affine space, with the inclusion-exclusion and product relations

$$[V \cup W] = [V] + [W] - [V \cap W]$$

$$[V] \cdot [W] = [V \times W].$$

By imposing equivalence under isomorphisms one falls back on the usual Grothendieck ring $K_0(\mathcal{V})$.

Thus, one can define an $\mathcal{R}$-valued algebro-geometric Feynman rule, for a given commutative ring $\mathcal{R}$, as in [Aluffi and Marcolli (2008b)] in terms of a ring homomorphism $I : \mathcal{F} \to \mathcal{R}$ by setting

$$\mathbb{U}(\Gamma) := I([\mathbb{A}^n]) - I([\hat{X}_\Gamma])$$

and by taking as value of the *inverse propagator*

$$\mathbb{U}(L) = I([\mathbb{A}^1]).$$

This then satisfies both (3.31) and (3.32). The ring $\mathcal{F}$ is then the receptacle of the universal algebro-geometric Feynman rule given by

$$\mathbb{U}(\Gamma) = [\mathbb{A}^n \smallsetminus \hat{X}_\Gamma] \in \mathcal{F}.$$

A Feynman rule defined in this way is *motivic* if the homomorphism I : $\mathcal{F} \to \mathcal{R}$ factors through the Grothendieck ring $K_0(\mathcal{V}_{\mathbb{K}})$. In this case the inverse propagator is the Lefschetz motive (or the propagator is the Tate motive).

The reason for working with $\mathcal{F}_{\mathbb{K}}$ instead is that it allowed us in [Aluffi and Marcolli (2008b)] to construct invariants of the graph hypersurfaces that behave like algebro-geometric Feynman rules and that measure to some extent how singular these varieties are, and which do not factor through the Grothendieck ring, since they contain specific information on how the $\hat{X}_\Gamma$ are embedded in the ambient affine space $\mathbb{A}^{\#E_{int}(\Gamma)}$.

3.9 Characteristic classes and Feynman rules

In the case of compact smooth varieties, the Chern classes of the tangent bundle can be written as a class $c(V) = c(TV) \cap [V]$ in homology whose degree of the zero dimensional component satisfies the Poincaré–Hopf theorem $\int c(TV) \cap [V] = \chi(V)$, which gives the topological Euler characteristic of the smooth variety.

Chern classes for singular varieties that generalize this property were introduced independently, following two different approaches, in [Schwartz (1965)] and [MacPherson (1974)]. The two definitions were later proved to be equivalent.

The approach followed by Marie Hélène Schwartz generalized the definition of Chern classes as the homology classes of the loci where a family of $k + 1$ vector fields become linearly dependent (for the lowest degree case one reads the Poincaré–Hopf theorem as saying that the Euler characteristic measures where a single vector field has zeros). In the case of singular varieties a generalization is obtained, provided that one assigns some radial conditions on the vector fields with respect to a stratification with good properties.

The approach followed by Robert MacPherson is instead based on functoriality. The functor $\mathbb{F}$ of constructible functions assigns to an algebraic variety V the $\mathbb{Z}$-module $\mathbb{F}(V)$ spanned by the characteristic functions 1_W of subvarieties $W \subset V$ and to a proper morphism $f : V \to V'$ the map $f_* : \mathbb{F}(V) \to \mathbb{F}(V')$ defined by $f_*(1_W)(p) = \chi(W \cap f^{-1}(p))$, with χ the Euler characteristic.

A conjecture of Grothendieck–Deligne predicted the existence of a unique natural transformation c_* between the functor $\mathbb{F}$ and the homol-

ogy (or Chow group) functor, which in the smooth case agrees with
$c_*(1_V) = c(TV) \cap [V]$.

MacPherson constructed this natural transformation in terms of data
of Mather classes and local Euler obstructions and defined in this way
what is usually referred to now as the Chern–Schwartz–MacPherson (CSM)
characteristic classes of (singular) algebraic varieties $c_{SM}(X) = c_*(1_X)$.

It is convenient here to work inside an ambient space, for instance an
ambient projective space, so that one can regard $c_*(1_X)$ as taking values
in the homology (or Chow group) of the ambient $\mathbb{P}^N$, so that the characteristic class $c_{SM}(X) = c_*(1_X)$ can be computed for X an arbitrary locally
closed subset of $\mathbb{P}^N$, without requiring a compactness hypothesis. This is
important in order to have an inclusion-exclusion relation for these classes,
as explained below.

The results of [Aluffi (2006)] show that, in fact, it is possible to compute
these classes without having to use Mather classes and Euler obstructions
that are usually very difficult to compute. Most notably, these characteristic
classes satisfy an inclusion-exclusion formula for a closed $Y \subset X$,

$$c_{SM}(X) = c_{SM}(Y) + c_{SM}(X \smallsetminus Y),$$

but are not invariant under isomorphism, hence they are naturally defined
on classes in $\mathcal{F}_{\mathbb{K}}$ but not on $K_0(\mathcal{V}_{\mathbb{K}})$.

For a smooth locally closed X in some ambient projective space $\mathbb{P}^N$,
the class $c_{SM}(X)$ can be obtained by a computation using Chern classes of
sheaves of differentials with logarithmic poles in a resolution of the closure
$\bar{X}$ of X. Then inclusion-exclusion gives a way to compute any embedded
c_{SM} without recourse to MacPherson's Euler obstructions or Chern-Mather
classes.

The CSM classes give good information on the singularities of a variety:
for example, in the case of hypersurfaces with isolated singularities, they
can be expressed in terms of Milnor numbers.

To construct a Feynman rule out of these Chern classes, one uses the
following procedure, [Aluffi and Marcolli (2008b)]. Given a variety $\hat{X} \subset \mathbb{A}^N$,
one can view it as a locally closed locus in $\mathbb{P}^N$, hence one can apply to
its characteristic function $1_{\hat{X}}$ the natural transformation c_* that gives an
element in the Chow group $A(\mathbb{P}^N)$ or in the homology $H_*(\mathbb{P}^N)$. This gives
as a result a class of the form

$$c_*(1_{\hat{X}}) = a_0[\mathbb{P}^0] + a_1[\mathbb{P}^1] + \cdots + a_N[\mathbb{P}^N].$$

One then defines an associated polynomial given by

$$G_{\hat{X}}(T) := a_0 + a_1 T + \cdots + a_N T^N.$$

It is in fact independent of N as it stops in degree equal to $\dim \hat{X}$. It is by construction invariant under linear changes of coordinates. It also satisfies an inclusion-exclusion property coming from the fact that the classes c_{SM} satisfy inclusion-exclusion, namely

$$G_{\hat{X} \cup \hat{Y}}(T) = G_{\hat{X}}(T) + G_{\hat{Y}}(T) - G_{\hat{X} \cap \hat{Y}}(T)$$

It is a more delicate result to show that it is multiplicative, namely that the following holds.

Theorem 3.9.1. *The polynomials $G_{\hat{X}}(T)$ satisfy*

$$G_{\hat{X} \times \hat{Y}}(T) = G_{\hat{X}}(T) \cdot G_{\hat{Y}}(T).$$

Proof. The proof of this fact is obtained in [Aluffi and Marcolli (2008b)] using an explicit formula for the CSM classes of joins in projective spaces, where the join $J(X, Y) \subset \mathbb{P}^{m+n-1}$ of two $X \subset \mathbb{P}^{m-1}$ and $Y \subset \mathbb{P}^{n-1}$ is defined as the set of

$$(sx_1 : \cdots : sx_m : ty_1 : \cdots : ty_n), \quad \text{with} \quad (s : t) \in \mathbb{P}^1,$$

and is related to the product in affine spaces by the property that the product $\hat{X} \times \hat{Y}$ of the affine cones over X and Y is the affine cone over $J(X, Y)$. One has

$$c_*(1_{J(X,Y)}) = ((f(H) + H^m)(g(H) + H^n) - H^{m+n}) \cap [\mathbb{P}^{m+n-1}],$$

where $c_*(1_X) = H^n f(H) \cap [\mathbb{P}^{n+m-1}]$ and $c_*(1_Y) = H^m g(H) \cap [\mathbb{P}^{n+m-1}]$, as functions of the hyperplane class H in $\mathbb{P}^{n+m-1}$. We refer the reader to [Aluffi and Marcolli (2008b)] for more details. $\square$

The resulting multiplicative property of the polynomials $G_{\hat{X}}(T)$ shows that one has a ring homomorphism $I_{CSM} : \mathcal{F} \to \mathbb{Z}[T]$ defined by

$$I_{CSM}([\hat{X}]) = G_{\hat{X}}(T)$$

and an associated Feynman rule

$$\mathbb{U}_{CSM}(\Gamma) = C_\Gamma(T) = I_{CSM}([\mathbb{A}^n]) - I_{CSM}([\hat{X}_\Gamma]).$$

This is not motivic, *i.e.* it does not factor through the Grothendieck ring $K_0(\mathcal{V}_{\mathbb{K}})$, as can be seen by the example given in [Aluffi and Marcolli (2008b)] of two graphs (see the figure below) that have different $\mathbb{U}_{CSM}(\Gamma)$,

$$C_{\Gamma_1}(T) = T(T+1)^2 \quad C_{\Gamma_2}(T) = T(T^2 + T + 1)$$

but the same hypersurface complement class in the Grothendieck ring,

$$[\mathbb{A}^n \setminus \hat{X}_{\Gamma_i}] = [\mathbb{A}^3] - [\mathbb{A}^2] \in K_0(\mathcal{V}).$$

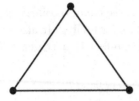 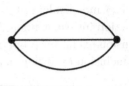

There is an interesting *positivity* property that the CSM classes of the graph hypersurfaces X_Γ appear to satisfy, namely the coefficients of all the powers H^k in the expression of the CMS class of X_Γ as a polynomial in H are positive. This was observed in [Aluffi and Marcolli (2008a)] on the basis of numerical computations of these classes, done using the program of [Aluffi (2003)], for sample graphs. As the graphs that are accessible to computer calculations of the corresponding CSM classes are necessarily combinatorially simple (and all planar), it is at present not known whether the observed positivity phenomenon holds more generally. This property may be related to other positivity phenomena observed for CSM classes of similarly combinatorial objects such as Schubert varieties [Aluffi and Mihalcea (2006)]. We see in the last chapter that, in the case of quantum field theories on noncommutative spacetimes and the corresponding modification of the graph polynomials, positivity fails in non-planar cases.

3.10   Deletion-contraction relation

We report here briefly the recent results of [Aluffi and Marcolli (2009b)] on deletion–contraction relations for motivic Feynman rules.

We have seen above that one can construct a polynomial invariant $C_\Gamma(T)$ of graphs that is an algebro-geometric Feynman rule in the sense of [Aluffi and Marcolli (2008b)], which does not factor through the Grothendieck ring of varieties but satisfies an inclusion–exclusion principle. The polynomial $C_\Gamma(T)$ is defined in terms of CSM classes of the graph hypersurface complement.

Several well known examples of polynomial invariants of graphs are examples of Tutte–Grothendieck invariants and are obtained as specializations of the Tutte polynomial of graphs, and they all have the property that they satisfy deletion–contraction relations, which make it possible to compute the invariant indictively from simpler graphs. These relations express the invariant of a given graph in terms of the invariants of the graphs obtained by deleting or contracting edges.

The results of [Aluffi and Marcolli (2009b)] investigate to what extent motivic and algebro-geometric Feynman rules satisfy deletion–contraction relations. Since such relations make it possible to control the invariant for more complicated graphs in terms of combinatorially simpler ones, the validity or failure of such relations can provide useful information that can help to identify where the motivic complexity of the graph hypersurfaces increases beyond the mixed Tate case (see §3.12 below).

In particular, it was shown in [Aluffi and Marcolli (2009b)] that the polynomial $C_\Gamma(T)$ is *not* a specialization of the Tutte polynomial. However, a form of deletion–contraction relation does hold for the motivic Feynman rule $[\mathbb{A}^n \smallsetminus \hat{X}_\Gamma] \in K_0(\mathcal{V})$. The latter, unlike the relation satisfied by Tutte–Grothendieck invariants, contains a term expressed as the class of an intersection of hypersurfaces which does not appear to be easily controllable in terms of combinatorial information alone. However, this form of deletion–contraction relations suffices to obtain explicit recursive relation for certain operations on graphs, such as replacing an edge by multiple parallel copies.

We start by recalling the case of the Tutte polynomial of graphs and the form of the deletion–contraction relation in that well known classical case. The Tutte polynomial $\mathcal{T}_\Gamma \in \mathbb{C}[x,y]$ of finite graphs Γ is completely determined by the following properties:

- If $e \in E(\Gamma)$ is neither a looping edge nor a bridge the deletion–contraction relation holds:

$$\mathcal{T}_\Gamma(x,y) = \mathcal{T}_{\Gamma \smallsetminus e}(x,y) + \mathcal{T}_{\Gamma/e}(x,y). \qquad (3.72)$$

- If $e \in E(\Gamma)$ is a looping edge then

$$\mathcal{T}_\Gamma(x,y) = y\mathcal{T}_{\Gamma/e}(x,y).$$

- If $e \in E(\Gamma)$ is a bridge then

$$\mathcal{T}_\Gamma(x,y) = x\mathcal{T}_{\Gamma \smallsetminus e}(x,y)$$

- If Γ has no edges then $\mathcal{T}_\Gamma(x,y) = 1$.

Here we call an edge e a "bridge" if the removal of e disconnects the graph Γ, and a "looping edge" if e starts and ends at the same vertex. Tutte–Grothendieck invariants are specializations of the Tutte polynomial. Among them one has well known invariants such as the chromatic polynomial and the Jones polynomial.

A first observation of [Aluffi and Marcolli (2009b)] is that the Tutte polynomial can be regarded as a Feynman rule, in the sense that it has the right multiplicative properties over disjoint unions and over decompositions of connected graphs in terms of trees and 1PI graphs.

Proposition 3.10.1. *The Tutte polynomial invariant defines an abstract Feynman rule with values in the polynomial ring $\mathbb{C}[x, y]$, by assigning*

$$U(\Gamma) = T_\Gamma(x, y), \quad \text{with inverse propagator} \quad U(L) = x. \quad (3.73)$$

Proof. One can first observe that the properties listed above for the Tutte polynomial determine the closed form

$$T_\Gamma(x, y) = \sum_{\gamma \subset \Gamma} (x - 1)^{\#V(\Gamma) - b_0(\Gamma) - (\#V(\gamma) - b_0(\gamma))} (y - 1)^{\#E(\gamma) - \#V(\gamma) + b_0(\gamma)},$$

$$(3.74)$$

where the sum is over subgraphs $\gamma \subset \Gamma$ with vertex set $V(\gamma) = V(\Gamma)$ and edge set $E(\gamma) \subset E(\Gamma)$. This can be written equivalently as

$$T_\Gamma(x, y) = \sum_{\gamma \subset \Gamma} (x - 1)^{b_0(\gamma) - b_0(\Gamma)} (y - 1)^{b_1(\gamma)}.$$

This is sometime referred to as the "sum over states" formula for the Tutte polynomial.

The multiplicative property under disjoint unions of graphs is clear from the closed expression (3.74), since for $\Gamma = \Gamma_1 \cup \Gamma_2$ we can identify subgraphs $\gamma \subset \Gamma$ with $V(\gamma) = V(\Gamma)$ and $E(\gamma) \subset E(\Gamma)$ with all possible pairs of subgraphs (γ_1, γ_2) with $V(\gamma_i) = V(\Gamma_i)$ and $E(\gamma_i) \subset E(\Gamma_i)$, with $b_0(\gamma) = b_0(\gamma_1) + b_0(\gamma_2)$, $\#V(\Gamma) = \#V(\Gamma_1) + \#V(\Gamma_2)$, and $\#E(\gamma) = \#E(\gamma_1) + \#E(\gamma_2)$. Thus, we get

$$T_\Gamma(x, y) = \sum_{\gamma = (\gamma_1, \gamma_2)} (x - 1)^{b_0(\gamma_1) + b_0(\gamma_2) - b_0(\Gamma)} (y - 1)^{b_1(\gamma_1) + b_1(\gamma_2)}$$

$$= T_{\Gamma_1}(x, y) \, T_{\Gamma_2}(x, y).$$

The property for connected and 1PI graphs follows from the fact that, when writing a connected graph in the form $\Gamma = \cup_{v \in V(T)} \Gamma_v$, with Γ_v 1PI graphs inserted at the vertices of the tree T, the internal edges of the tree are all bridges in the resulting graph, hence the property of the Tutte polynomial for the removal of bridges gives

$$T_\Gamma(x, y) = x^{\#E_{int}(T)} \, T_{\Gamma \smallsetminus \cup_{e \in E_{int}(T)} e}(x, y).$$

Then one obtains an abstract Feynman rule with values in $\mathcal{R} = \mathbb{C}[x, y]$ of the form (3.73). $\qquad \square$

It follows that Tutte–Grothendieck invariants such as the chromatic and Jones polynomials can also be regarded as Feynman rules.

One can then observe, however, that the algebro–geometric Feynman rule

$$C_\Gamma(T) = I_{CSM}([\mathbb{A}^n \smallsetminus \hat{X}_\Gamma])$$

that we discussed in the previous section is not a specialization of the Tutte polynomial.

Proposition 3.10.2. *The polynomial invariant $C_\Gamma(T)$ is not a specialization of the Tutte polynomial.*

Proof. We show that one cannot find functions $x = x(T)$ and $y = y(T)$ such that

$$C_\Gamma(T) = T_\Gamma(x(T), y(T)).$$

First notice that, if $e \in E(\Gamma)$ is a bridge, the polynomial $C_\Gamma(T)$ satisfies the relation

$$C_\Gamma(T) = (T+1)C_{\Gamma \smallsetminus e}(T). \tag{3.75}$$

In fact, $(T+1)$ is the inverse propagator of the algebro-geometric Feynman rule $U(\Gamma) = C_\Gamma(T)$ and the property of abstract Feynman rules for 1PI graphs connected by a bridge gives (3.75). In the case where $e \in E(\Gamma)$ is a looping edge, we have

$$C_\Gamma(T) = T \, C_{\Gamma/e}(T). \tag{3.76}$$

In fact, adding a looping edge to a graph corresponds, in terms of graph hypersurfaces, to taking a cone on the graph hypersurface and intersecting it with the hyperplane defined by the coordinate of the looping edge. This implies that the universal algebro-geometric Feynman rule with values in the Grothendieck ring $\mathcal{F}$ of immersed conical varieties satisfies

$$U(\Gamma) = ([\mathbb{A}^1] - 1)U(\Gamma/e)$$

if e is a looping edge of Γ and $U(\Gamma) = [\mathbb{A}^n \smallsetminus \hat{X}_\Gamma] \in \mathcal{F}$. The property (3.76) then follows since the image of the class $[\mathbb{A}^1]$ is the inverse propagator $(T+1)$. (See §3.9 above and Proposition 2.5 and §2.2 of [Aluffi and Marcolli (2008b)].)

This implies that, if $C_\Gamma(T)$ has to be a specialization of the Tutte polynomial, the relations for bridges and looping edges imply that one has to identify $x(T) = T + 1$ and $y(T) = T$. However, this is not compatible with the behavior of the invariant $C_\Gamma(T)$ on more complicated graphs. For example, for the triangle graph one has $C_\Gamma(T) = T(T+1)^2$ while the specialization $T_\Gamma(x(T), y(T)) = (T+1)^2 + (T+1) + T$. $\quad\square$

One can then investigate what kind of deletion–contraction relations hold in the case of motivic and algebro–geometric Feynman rules. An answer is given in [Aluffi and Marcolli (2009b)] for the motivic case, that is, for the Feynman rule

$$\mathbb{U}(\Gamma) = [\mathbb{A}^n \smallsetminus \hat{X}_\Gamma] \in K_0(\mathcal{V}).$$

Theorem 3.10.3. *Let Γ be a graph with $n > 1$ edges.*

(1) If $e \in E(\Gamma)$ is neither a bridge nor a looping edge then the following deletion–contraction relation holds:

$$[\mathbb{A}^n \smallsetminus \hat{X}_\Gamma] = \mathbb{L} \cdot [\mathbb{A}^{n-1} \smallsetminus (\hat{X}_{\Gamma \smallsetminus e} \cap \hat{X}_{\Gamma/e})] - [\mathbb{A}^{n-1} \smallsetminus \hat{X}_{\Gamma \smallsetminus e}]. \quad (3.77)$$

(2) If the edge e is a bridge in Γ, then

$$[\mathbb{A}^n \smallsetminus \hat{X}_\Gamma] = \mathbb{L} \cdot [\mathbb{A}^{n-1} \smallsetminus \hat{X}_{\Gamma \smallsetminus e}] = \mathbb{L} \cdot [\mathbb{A}^{n-1} \smallsetminus \hat{X}_{\Gamma/e}]. \quad (3.78)$$

(3) If e is a looping edge in Γ, then

$$[\mathbb{A}^n \smallsetminus \hat{X}_\Gamma] = (\mathbb{L} - 1) \cdot [\mathbb{A}^{n-1} \smallsetminus \hat{X}_{\Gamma \smallsetminus e}] = (\mathbb{L} - 1) \cdot [\mathbb{A}^{n-1} \smallsetminus \hat{X}_{\Gamma/e}]. \quad (3.79)$$

(4) If Γ contains at least two loops, then the projective version of (3.77) also holds, in the form

$$[\mathbb{P}^{n-1} \smallsetminus X_\Gamma] = \mathbb{L} \cdot [\mathbb{P}^{n-2} \smallsetminus (X_{\Gamma \smallsetminus e} \cap X_{\Gamma/e})] - [\mathbb{P}^{n-2} \smallsetminus X_{\Gamma \smallsetminus e}]. \quad (3.80)$$

(5) Under the same hypotheses, the Euler characteristics satisfy

$$\chi(X_\Gamma) = n + \chi(X_{\Gamma \smallsetminus e} \cap X_{\Gamma/e}) - \chi(X_{\Gamma \smallsetminus e}). \quad (3.81)$$

Proof. (1) and (4): Let Γ be a graph with $n \geq 2$ edges $e_1, \ldots, e_{n-1}, e = e_n$, with $(t_1 : \ldots : t_n)$ the corresponding variables in $\mathbb{P}^{n-1}$. As above we consider the Kirchhoff polynomial Ψ_Γ and the affine and projective graph hypersurfaces $\hat{X}_\Gamma \subset \mathbb{A}^n$ and $X_\Gamma \subset \mathbb{P}^{n-1}$. We assume $\deg \Psi_\Gamma = \ell > 0$.

We consider an edge e that is not a bridge or looping edge. Then the polynomials for the deletion $\Gamma \smallsetminus e$ and contraction Γ/e of the edge $e = e_n$ are both non-trivial and given by

$$F := \frac{\partial \Psi_\Gamma}{\partial t_n} = \Psi_{\Gamma \smallsetminus e} \quad \text{and} \quad G := \Psi_\Gamma|_{t_n=0} = \Psi_{\Gamma/e}. \quad (3.82)$$

The Kirchhoff polynomial Ψ_Γ satisfies

$$\Psi_\Gamma(t_1, \ldots, t_n) = t_n F(t_1, \ldots, t_{n-1}) + G(t_1, \ldots, t_{n-1}). \quad (3.83)$$

Given a projective hypersurface $Y \subset \mathbb{P}^{N-1}$ we denote by $\hat{Y}$ the affine cone in $\mathbb{A}^N$, and we use the notation $\bar{Y}$ for the projective cone in $\mathbb{P}^N$.

It is proved in [Aluffi and Marcolli (2009b)], Theorem 3.3, in a more general context that includes the case under consideration, that the projection from the point $(0 : \cdots : 0 : 1)$ induces an isomorphism

$$X_\Gamma \smallsetminus (X_\Gamma \cap \overline{X}_{\Gamma \smallsetminus e}) \xrightarrow{\sim} \mathbb{P}^{n-2} \smallsetminus X_{\Gamma \smallsetminus e}, \qquad (3.84)$$

where $X_{\Gamma \smallsetminus e}$ is the hypersurface in $\mathbb{P}^{n-2}$ defined by the polynomial F. In fact, the projection $\mathbb{P}^{n-1} \dashrightarrow \mathbb{P}^{n-2}$ from $p = (0 : \cdots : 0 : 1)$ acts as

$$(t_1 : \cdots : t_n) \mapsto (t_1 : \cdots : t_{n-1}).$$

If F is constant (that is, if $\deg \psi = 1$), then $X_{\Gamma \smallsetminus e} = \overline{X}_{\Gamma \smallsetminus e} = \emptyset$ and the statement is trivial. Thus, assume $\deg F > 0$. In this case, $\Psi_\Gamma(p) = F(p) = 0$, hence $p \in X_\Gamma \cap \overline{X}_{\Gamma \smallsetminus e}$, and hence $p \notin X \smallsetminus (X_\Gamma \cap \overline{X}_{\Gamma \smallsetminus e})$. Therefore, the projection restricts to a regular map

$$X_\Gamma \smallsetminus (X_\Gamma \cap \overline{X}_{\Gamma \smallsetminus e}) \to \mathbb{P}^{n-2}.$$

The image is clearly contained in $\mathbb{P}^{n-2} \smallsetminus X_{\Gamma \smallsetminus e}$, and the statement is that this map induces an *isomorphism*

$$X_\Gamma \smallsetminus (X_\Gamma \cap \overline{X}_{\Gamma \smallsetminus e}) \xrightarrow{\sim} \mathbb{P}^{n-2} \smallsetminus X_{\Gamma \smallsetminus e} \quad.$$

Let $q = (q_1 : \cdots : q_{n-1})$. The line from p to q is parametrized by

$$(q_1 : \cdots : q_{n-1} : t).$$

We show that this line meets $X_\Gamma \smallsetminus (X_\Gamma \cap \overline{X}_{\Gamma \smallsetminus e})$ transversely at one point. The intersection with X_Γ is determined by the equation

$$t F(q_1 : \cdots : q_{n-1}) + G(q_1 : \cdots : q_{n-1}) = 0.$$

Since $F(q) \neq 0$, this is a polynomial of degree exactly 1 in t, and determines a reduced point, as needed.

It then follows that, in the Grothendieck ring $K_0(\mathcal{V})$ we have identities

$$[\mathbb{P}^{n-1} \smallsetminus X_\Gamma] = [\mathbb{P}^{n-1} \smallsetminus (X_\Gamma \cap \overline{X}_{\Gamma \smallsetminus e})] - [\mathbb{P}^{n-2} \smallsetminus X_{\Gamma \smallsetminus e}]. \qquad (3.85)$$

If Γ has at least two loops, so that $\deg \Psi_\Gamma > 1$, then we also have

$$[\mathbb{P}^{n-1} \smallsetminus X_\Gamma] = \mathbb{L} \cdot [\mathbb{P}^{n-2} \smallsetminus (X_{\Gamma \smallsetminus e} \cap X_{\Gamma/e})] - [\mathbb{P}^{n-2} \smallsetminus X_{\Gamma \smallsetminus e}], \qquad (3.86)$$

where $\mathbb{L} = [\mathbb{A}^1]$ is the Lefschetz motive and $X_{\Gamma/e}$ is the hypersurface $G = 0$. Indeed, the ideal of $X_\Gamma \cap \overline{X}_{\Gamma \smallsetminus e}$ is

$$(\psi, F) = (t_n F + G, F) = (F, G).$$

This means that

$$X_\Gamma \cap \overline{X}_{\Gamma \smallsetminus e} = \overline{X}_{\Gamma \smallsetminus e} \cap \overline{X}_{\Gamma/e}. \qquad (3.87)$$

If $\deg X_\Gamma > 1$, then F is not constant, hence $\overline{X}_{\Gamma \smallsetminus e} \neq \emptyset$. It then follows that $\overline{X}_{\Gamma \smallsetminus e} \cap \overline{X}_{\Gamma/e}$ contains the point $p = (0 : \cdots : 0 : 1)$. The fibers of the projection

$$\mathbb{P}^{n-1} \smallsetminus (\overline{X}_{\Gamma \smallsetminus e} \cap \overline{X}_{\Gamma/e}) \to \mathbb{P}^{n-2} \smallsetminus X_{\Gamma \smallsetminus e}$$

with center p are then all isomorphic to $\mathbb{A}^1$, and it follows that

$$[\mathbb{P}^{n-1} \smallsetminus (\overline{X}_{\Gamma \smallsetminus e} \cap \overline{X}_{\Gamma/e})] = \mathbb{L} \cdot [\mathbb{P}^{n-2} \smallsetminus (X_{\Gamma \smallsetminus e} \cap X_{\Gamma/e})].$$

The affine version (3.77) follows by observing that, if $\deg X_\Gamma > 1$, then $\deg F > 0$, hence

$$\widehat{X_\Gamma \cap \overline{X}_{\Gamma \smallsetminus e}} \quad \text{and} \quad \widehat{X}_{\Gamma \smallsetminus e} \cap \widehat{X}_{\Gamma/e}$$

contain the origin. In this case, the classes in the Grothendieck ring satisfy

$$[\mathbb{A}^N \smallsetminus \widehat{X_\Gamma \cap \overline{X}_{\Gamma \smallsetminus e}}] = (\mathbb{L} - 1) \cdot [\mathbb{P}^{N-1} \smallsetminus (X_\Gamma \cap \overline{X}_{\Gamma \smallsetminus e})]$$

$$[\mathbb{A}^N \smallsetminus (\widehat{X}_{\Gamma \smallsetminus e} \cap \widehat{X}_{\Gamma/e})] = (\mathbb{L} - 1) \cdot [\mathbb{P}^{N-1} \smallsetminus (X_{\Gamma \smallsetminus e} \cap X_{\Gamma/e})].$$

Then (3.77) follows from the formula for the projective case, while it can be checked directly for the remaning case with $\deg X_\Gamma = 1$.

(2) If e is a bridge, then Ψ_Γ does not depend on the variable t_e and $F \equiv 0$. The equation for $X_{\Gamma \smallsetminus e}$ is $\Psi_\Gamma = 0$ again, but viewed in one fewer variables. The equation for $X_{\Gamma/e}$ is the same.

(3) If e is a looping edge, then Ψ_Γ is divisible by t_e, so that $G \equiv 0$. The equation for $X_{\Gamma/e}$ is obtained by dividing Ψ_Γ through by t_e, and one has $X_{\Gamma \smallsetminus e} = X_{\Gamma/e}$.

(5) Under the hypothesis that $\deg X_\Gamma > 1$, the statement about the Euler characteristics follows from (3.80). One has from (3.84) the identity

$$\chi(\mathbb{P}^{n-1} \smallsetminus (X_\Gamma \cup \overline{X}_{\Gamma \smallsetminus e})) = 0.$$

In the case where $\deg X_\Gamma = 1$ one has $\chi(X_\Gamma) = n - 1$. $\qquad\square$

This general result on deletion-contraction relations for motivic Feynman rules was applied in [Aluffi and Marcolli (2009b)] to analyze some operations on graphs, which have the property that the problem of describing the intersection $X_{\Gamma \smallsetminus e} \cap X_{\Gamma/e}$ can be bypassed and the class of more complicated graphs can be computed inductively only in terms of combinatorial data.

One such operation replaces a chosen edge e in a graph Γ with m parallel edges connecting the same two vertices $\partial(e)$. Another operation consists of replacing an edge with a chain of triangles (referred to as a "lemon graph").

The latter operation can easily be generalized to chains of polygons as explained in [Aluffi and Marcolli (2009b)].

Assume that e is an edge of Γ, and denote by Γ_{me} the graph obtained from Γ by replacing e by m parallel edges. (Thus, $\Gamma_{0e} = \Gamma \smallsetminus e$, and $\Gamma_e = \Gamma$.)

Theorem 3.10.4. *Let e be an edge of a graph Γ. Let $\mathbb{T} = \mathbb{L} - 1 = [\mathbb{G}_m] \in K_0(\mathcal{V})$.*

(1) If e is a looping edge, then

$$\sum_{m \geq 0} \mathbb{U}(\Gamma_{me}) \frac{s^m}{m!} = e^{\mathbb{T}s} \, \mathbb{U}(\Gamma \smallsetminus e). \tag{3.88}$$

(2) If e is a bridge, then

$$\sum_{m \geq 0} \mathbb{U}(\Gamma_{me}) \frac{s^m}{m!} = \left(\mathbb{T} \cdot \frac{e^{\mathbb{T}s} - e^{-s}}{\mathbb{T} + 1} + s \, e^{\mathbb{T}s} + 1 \right) \mathbb{U}(\Gamma \smallsetminus e). \tag{3.89}$$

(3) If e is not a bridge nor a looping edge, then

$$\sum_{m \geq 0} \mathbb{U}(\Gamma_{me}) \frac{s^m}{m!} = \frac{e^{\mathbb{T}s} - e^{-s}}{\mathbb{T} + 1} \mathbb{U}(\Gamma)$$

$$+ \frac{e^{\mathbb{T}s} + \mathbb{T}e^{-s}}{\mathbb{T} + 1} \mathbb{U}(\Gamma \smallsetminus e) \tag{3.90}$$

$$+ \left(s \, e^{\mathbb{T}s} - \frac{e^{\mathbb{T}s} - e^{-s}}{\mathbb{T} + 1} \right) \mathbb{U}(\Gamma/e).$$

The result is proved in [Aluffi and Marcolli (2009b)] by first deriving the relation in the case of doubling an edge and then repeating the construction. We do not report here the details of the proof, and we refer the readers to the paper for more details, but it is worth pointing out that the main step that makes it possible to obtain in this case a completely explicit formula in terms of $\mathbb{U}(\Gamma)$, $\mathbb{U}(\Gamma \smallsetminus e)$ and $\mathbb{U}(\Gamma/e)$ is a cancellation that occurs in the calculation of the classes in the operation of doubling an edge. In fact, one first expresses the class $\mathbb{U}(\Gamma_{2e})$ as

$$\mathbb{U}(\Gamma_{2e}) = \mathbb{L} \cdot [\mathbb{A}^n \smallsetminus (\hat{X}_\Gamma \cap \hat{X}_{\Gamma_o})] - \mathbb{U}(\Gamma),$$

where $n = \#E(\Gamma)$ and Γ_o denotes the graph obtained by attaching a looping edge named e to Γ/e. In the notation we used earlier in this section, the equation for Γ_o is given by the polynomial $t_e \, G$. Using inclusion–exclusion for classes in the Grothendieck ring, it is then shown in [Aluffi and Marcolli (2009b)] that one has

$$[\hat{X}_\Gamma \cap \hat{X}_{\Gamma_o}] = [\hat{X}_{\Gamma/e}] + (\mathbb{L} - 1) \cdot [\hat{X}_{\Gamma \smallsetminus e} \cap \hat{X}_{\Gamma/e}].$$

Thus, one obtains

$$\mathbb{U}(\Gamma_{2e}) = (\mathbb{L} - 2) \cdot \mathbb{U}(\Gamma) + (\mathbb{L} - 1) \cdot \mathbb{U}(\Gamma \smallsetminus e) + \mathbb{L} \cdot \mathbb{U}(\Gamma/e).$$

This expression no longer contains classes of intersections of graph hypersurfaces, and can be iterated to obtain the general formula for $\mathbb{U}(\Gamma_{me})$.

It is shown in [Aluffi and Marcolli (2009b)] that the recursive relation obtained in this way for the operation of multiplying edges is very similar in form to the recursive relation that the Tutte polynomial satisfies under the same operation. In fact, they both are solutions to a universal recursive relation with different initial conditions. One also derives from the same universal recursion a conjectural form for the recursion relation, under the operation of multiplying edges $\Gamma \mapsto \Gamma_{me}$, for the polynomial invariant $C_\Gamma(T) = I_{CSM}([\mathbb{A}^n \smallsetminus \hat{X}_\Gamma])$.

The operation of replacing an edge with a chain of triangles, or "lemon graph" depicted in the figure produces, similarly, a recursive formula for the motivic Feynman rule $\mathbb{U}(\Gamma_m^\Lambda)$, where Γ_m^Λ is the graph obtained from Γ by replacing an edge e by a lemon graph Λ_m.

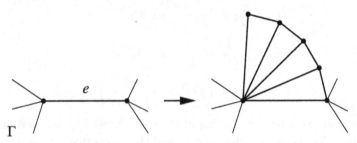

Proposition 3.10.5. *Let e be an edge of a graph Γ, and assume that e is neither a bridge nor a looping edge. Let Γ_m^Λ be the "lemonade graph" obtained by building an m-lemon fanning out from e. Then*

$$\sum_{m \geq 0} \mathbb{U}(\Gamma_m^\Lambda) s^m = \frac{1}{1 - \mathbb{T}(\mathbb{T} + 1)s - \mathbb{T}(\mathbb{T} + 1)^2 s^2}$$
$$\cdot \left((1 - (\mathbb{T} + 1)s) \mathbb{U}(\Gamma) + (\mathbb{T} + 1)\mathbb{T}s \, \mathbb{U}(\Gamma \smallsetminus e) + (\mathbb{T} + 1)^2 s \, \mathbb{U}(\Gamma/e)\right).$$

This result is proven in [Aluffi and Marcolli (2009b)] by first obtaining a recursive formula for the classes $\mathbb{U}(\Lambda_m)$ of the lemon graphs. This is a succession of operations where one first doubles an edge and then splits the added edge by inserting a vertex. The first operation is covered by the expression derived before for the class $\mathbb{U}(\Gamma_{2e})$ while splitting an edge with a vertex corresponds to taking a cone over the corresponding graph hypersurface. Thus, both operations are well understood at the level of classes in the Grothendieck ring and they give the recursive relation

$$\mathbb{U}(\Lambda_{m+1}) = \mathbb{T}(\mathbb{T}+1)\mathbb{U}(\Lambda_m) + \mathbb{T}(\mathbb{T}+1)^2\mathbb{U}(\Lambda_{m-1}),$$

for $m \geq 1$, which gives then

$$\mathbb{U}(\Lambda_m) = (\mathbb{T}+1)^{m+1} \sum_{i=0}^{m} \binom{m-i}{i} \mathbb{T}^{m-i}.$$

It is interesting to observe that the sequence $a_m = \mathbb{U}(\Lambda_m)$ is a divisibility sequence, in the sense that $\mathbb{U}(\Lambda_{m-1})$ divides $\mathbb{U}(\Lambda_{n-1})$ if m divides n. This is the property satisfied, for instance, by the sequence of Fibonacci numbers. (See [Aluffi and Marcolli (2009b)], Corollary 5.12 for more details.)

Returning to the classes $\mathbb{U}(\Gamma_m^\Lambda)$, one then shows that these also satisfy a relation of the form

$$\mathbb{U}(\Gamma_m^\Lambda) = f_m(\mathbb{T})\mathbb{U}(\Gamma) + g_m(\mathbb{T})\mathbb{U}(\Gamma \smallsetminus e) + h_m(\mathbb{T})\mathbb{U}(\Gamma/e),$$

where the f_m, g_m, h_m satisfy the same recursion that gives $\mathbb{U}(\Lambda_m)$, but with different seeds given by

$$\begin{cases} f_0(\mathbb{T}) = 1 &, \quad f_1(\mathbb{T}) = \mathbb{T}^2 - 1 \\ g_0(\mathbb{T}) = 0 &, \quad g_1(\mathbb{T}) = \mathbb{T}(\mathbb{T}+1) \\ h_0(\mathbb{T}) = 0 &, \quad h_1(\mathbb{T}) = (\mathbb{T}+1)^2. \end{cases}$$

For more details we refer the reader to [Aluffi and Marcolli (2009b)].

3.11 Feynman integrals and periods

Numerical calculations performed in [Broadhurst and Kreimer (1997)] gave very strong evidence for a relation between residues of Feynman integrals and periods of mixed Tate motives. In fact, the numerical evidence indicates that the values computed in the log divergent case are $\mathbb{Q}$-linear combinations of multiple zeta values, which, as we recalled in the previous chapter, can be realized as periods of mixed Tate motives.

Multiple zeta values are real numbers obtained by summing convergent series of the form

$$\zeta(n_1, \ldots, n_r) = \sum_{0 < k_1 < k_2 < \cdots < k_r} \frac{1}{k_1^{n_1} \cdots k_r^{n_r}}, \tag{3.91}$$

where the n_i are positive integers with $n_r \geq 2$. There are ways to realize multiple zeta values as periods of mixed Tate motives; see [Goncharov

(2001)], [Goncharov and Manin (2004)], [Terasoma (2002)]. In [Goncharov and Manin (2004)], for instance, as well as in [Brown (2006)], multiple zeta values are explicitly realized as periods on the moduli spaces of curves. It is the occurrence of multiple zeta values in Feynman integral computations, along with the fact that these numbers arise as periods of mixed Tate motives, that gave rise to the idea of a deeper relation between perturbative quantum field theory and motives.

Among the examples of Feynman graphs computed in [Broadhurst and Kreimer (1997)], the wheel with n spokes graphs give simple zeta values $\zeta(2n - 3)$. It is also shown in [Broadhurst and Kreimer (1997)] that the non-planar graph given by the complete bipartite graph $K_{3,4}$ evaluates to the double zeta value $\zeta(5, 3)$.

Recent results of [Brown (2009a)] for multiloop Feynman diagrams provide a new method for evaluating the case of primitive divergent graphs (*i.e.* those that do not contain any smaller divergent subgraph, and are therefore primitive elements in the Connes–Kreimer Hopf algebra). This shows the occurrence of values at roots of unity of multiple polylogarithms, in addition to the multiple zeta values observed in [Broadhurst and Kreimer (1997)]. The multiple polylogarithm function, introduced in [Goncharov (2001)] is defined as

$$\mathrm{Li}_{n_1,\ldots,n_r}(x_1,\ldots,x_r) = \sum_{0<k_1<k_2<\cdots<k_r} \frac{x_1^{k_1}\cdots x_r^{k_r}}{k_1^{n_1}\cdots k_r^{n_r}}. \tag{3.92}$$

It is absolutely convergent for $|x_i| < 1$ and it extends analytically to an open domain in $\mathbb{C}^r$ as a multivalued holomorphic function. As shown in [Goncharov (2001)], the multiple polylogarithms are also associated to objects in the category of mixed Tate motives.

3.12   The mixed Tate mystery

The results recalled in the previous section suggest that, in its strongest possible form, one could formulate a conjecture as follows.

Conjecture 3.12.1. *Are all residues of Feynman graphs of perturbative scalar field theories (possibly after a suitable regularization and renormalization procedure is used) periods of mixed Tate motives?*

In this form the conjecture is somewhat vaguely formulated, but it does not matter since it is unlikely to hold true in such generality. One can however refine the question and formulate it more precisely in the following form.

Question 3.12.2. *Under what conditions on the graphs, the scalar field theory, and the renormalization procedure are the residues of Feynman graphs periods of mixed Tate motives?*

As we have seen above, using the Feynman parameters, one can write the Feynman integrals in a form that looks exactly like a period of an algebraic variety obtained as a hypersurface complement, modulo the issue of divergences which needs to be handled separately. In the stable range for D, this hypersurface is the graph variety X_Γ, while in the unstable range it is the Landau variety Y_Γ, or the union of the two.

Kontsevich formulated the conjecture that the graph hypersurfaces X_Γ themselves may always be mixed Tate motives, which would imply Conjecture 3.12.1. Although numerically this conjecture was at first verified up to a large number of loops in [Stembridge (1998)], it was later disproved in [Belkale and Brosnan (2003a)]. They proved that the varieties X_Γ can be arbitrarily complicated as motives: indeed, the X_Γ generate the Grothendieck ring of varieties.

The fact that the classes $[X_\Gamma]$ can be arbitrarily far from the mixed Tate part $\mathbb{Z}[\mathbb{L}]$ of the Grothendieck ring $K_0(\mathcal{V})$ seems to indicate at first that Conjecture 3.12.1 cannot hold. However, that is not necessarily the case. In fact, only a part of the cohomology of the complement of the hypersurface X_Γ is involved in the period computation of the Feynman integral. As we have seen earlier in this chapter, ignoring divergences, one is in fact considering only a certain relative cohomology group, namely

$$H^{n-1}(\mathbb{P}^{n-1} \setminus X_\Gamma, \Sigma_n \setminus (\Sigma_n \cap X_\Gamma)), \qquad (3.93)$$

where $n = \#E_{int}(\Gamma)$ and $\Sigma_n = \{t \in \mathbb{P}^{n-1} \mid \prod_i t_i = 0\}$ denotes the normal crossings divisor given by the union of the coordinate hyperplanes. In fact, if one ignores momentarily the issue of divergences, the evaluation of the integral

$$\int_{\sigma_n} \frac{P_\Gamma(p,t)^{-n+D\ell/2}}{\Psi_\Gamma(t)^{-n+(\ell+1)D/2}} \, \omega_n \qquad (3.94)$$

can be seen as a pairing of a differential form on the hypersurface complement with a chain σ_n with boundary $\partial\sigma_n \subset \Sigma_n$ contained in the normal crossings divisor. Thus, one is working with a relative cohomology, and we have seen that the question of whether the integral above (*i.e.* the residue of the Feynman graph after dropping the divergent Gamma factor in the parametric Feynman integral) evaluates to a period of a mixed Tate motive can be addressed in terms of the question of whether the relative cohomology (3.93) is a realization of a mixed Tate motive

$$\mathfrak{m}(\mathbb{P}^{n-1} \smallsetminus X_\Gamma, \Sigma_n \smallsetminus X_\Gamma \cap \Sigma_n). \tag{3.95}$$

This could still be the case even though the motive of X_Γ itself need not be mixed Tate. However, this observation makes it clear that, if something like Conjecture 3.12.1 holds, it is due to a very subtle interplay between the geometry of the graph hypersurface X_Γ and the locus of intersection $X_\Gamma \cap \Sigma_n$. Notice that this locus is also the one that is responsible for the divergences of the integral (3.94), so that the situation is further complicated, when one tries to eliminate divergences, by possibly performing blow-ups of this locus. Even without introducing immediately the further difficulties related to eliminating divergences, the main problem remains that as the graphs increase in combinatorial complexity, it becomes extemely difficult to control the motivic nature of (3.95). Thus, at present, not only is the conjecture in its stronger form 3.12.1 still open, but not much is understood on specific conditions that would ensure that (3.95) is in fact mixed Tate, according to the formulation of Question 3.12.2.

Some new work of Francis Brown, which became available in the preprint [Brown (2009b)] while this book was going to print, identifies specific families of graphs for which (3.95) is shown to give a mixed Tate motive and for which the corresponding Feynman integrals evaluate to multiple zeta values.

In fact, a weaker statement than the relative cohomology (3.93) being mixed Tate may still suffice to prove that the residue of the Feynman integral is a period of a mixed Tate motive, since the period only captures a piece of the relative cohomology (3.93), so that the question on the motivic nature of this relative cohomology provides a sufficient condition for a mixed Tate period, but not a necessary one.

Very nice recent results of [Doryn (2008)] compute the middle cohomology of the graph hypersurfaces for a larger class of graphs than those originally considered in the work of [Bloch, Esnault, Kreimer (2006)], consisting of so called "zig-zag graphs". This identifies in a significant class of examples a mixed Tate piece of the cohomology.

As we mentioned in relation to the recent result of [Bloch (2008)] on sums of graphs with fixed number of vertices, one should also take into account that the possible occurrence of non-mixed Tate periods in the Feynman amplitudes for more complicated graphs may cancel out in sums over graphs with fixed external structure or fixed number of vertices and leave only a mixed Tate contribution.

One should also keep in mind that, in the case of divergent Feynman integrals, handling divergences in different ways might affect the nature of the resulting periods, after subtraction of infinities.

3.13 From graph hypersurfaces to determinant hypersurfaces

An attempt to provide specific conditions, as stated in Question 3.12.2 above, that would ensure that certain Feynman integrals evaluate to periods of mixed Tate motives was developed in [Aluffi and Marcolli (2009a)].

One can use the properties of periods, in particular the change of variable formula, to recast the computation of a given integral $\int_\sigma \omega$ computing a period in a different geometric ambient variety, whose motivic nature is easier to control. As long as no information is lost in the period computation, one can hope to obtain in this way some sufficient conditions that will ensure that the periods associated to Feynman integrals are mixed Tate.

A period $\int_\sigma \omega$ associated to the data (X, D, ω, σ) of a variety X, a divisor D, a differential form ω on X, and an integration domain σ with boundary $\partial\sigma \subset D$ can be computed equivalently after a change of variables obtained by mapping it via a morphism f of varieties to another set of data $(X', D', \omega', \sigma')$, with $\omega = f^*(\omega')$ and $\sigma' = f_*(\sigma)$.

In the case of parametric Feynman integrals one can proceed in the following way. The matrix $M_\Gamma(t)$ associated to a Feynman graph Γ determines a linear map of affine spaces

$$\Upsilon : \mathbb{A}^n \to \mathbb{A}^{\ell^2}, \quad \Upsilon(t)_{kr} = \sum_i t_i \eta_{ik} \eta_{ir} \tag{3.96}$$

such that the affine graph hypersurface is obtained as the preimage

$$\hat{X}_\Gamma = \Upsilon^{-1}(\hat{\mathcal{D}}_\ell)$$

under this map of the determinant hypersurface,

$$\hat{\mathcal{D}}_\ell = \{x = (x_{ij}) \in \mathbb{A}^{\ell^2} \mid \det(x_{ij}) = 0\}.$$

One can give explicit combinatorial conditions on the graph that ensure that the map Υ is an embedding. As shown in [Aluffi and Marcolli (2009a)], for any 3-edge-connected graph with at least 3 vertices and no looping edges, which admit a closed 2-cell embedding of face width at least 3, the map Υ

is injective. To see why this is the case, one can start with the following observation on the properties of the matrix $M_\Gamma(t)$ that defines the map Υ of (3.96).

Lemma 3.13.1. *The matrix* $M_\Gamma(t) = \eta^\dagger \Lambda \eta$ *defining the map* Υ *has the following properties.*

- *For* $i \neq j$, *the corresponding entry is the sum of* $\pm t_k$, *where the* t_k *correspond to the edges common to the* i-*th and* j-*th loop, and the sign is* $+1$ *if the orientations of the edges both agree or both disagree with the loop orientations, and* -1 *otherwise.*
- *For* $i = j$, *the entry is the sum of the variables* t_k *corresponding to the edges in the* i-*th loop (all taken with sign* $+$).

Similarly, for a specific edge e, *with* t_e *the corresponding variable, one has the following.*

- *The variable* t_e *appears in* $\eta^\dagger \Lambda \eta$ *if and only if* e *is part of at least one loop.*
- *If* e *belongs to a single loop* ℓ_i, *then* t_e *only appears in the diagonal entry* (i, i), *added to the variables corresponding to the other edges forming the loop* ℓ_i.
- *If there are two loops* ℓ_i, ℓ_j *containing* e, *and not having any other edge in common, then the* $\pm t_e$ *appears by itself at the entries* (i, j) *and* (j, i) *in the matrix* $\eta^\dagger \Lambda \eta$.

In the following, we denote by Υ_i the composition of the map Υ with the projection to the i-th row of the matrix $\eta^\dagger \Lambda \eta$, viewed as a map of the variables corresponding only to the edges that belong to the i-th loop in the chosen bases of the first homology of the graph Γ.

Lemma 3.13.2. *If* Υ_i *is injective for* i *ranging over a set of loops such that every edge of* Γ *is part of a loop in that set, then* Υ *is itself injective.*

Proof. Let $(t_1, \ldots, t_n) = (c_1, \ldots, c_n)$ be in the kernel of τ. Since each (i, j) entry in the target matrix is a combination of edges in the i-th loop, the map τ_i must send to zero the tuple of c_j's corresponding to the edges in the i-th loop. Since we are assuming τ_i to be injective, that tuple is the zero-tuple. Since every edge is in some loop for which τ_i is injective, it follows that every c_j is zero, as needed. $\square$

One can then give conditions based on properties of the graph that ensure that the components Υ_i are injective. This is done in [Aluffi and Marcolli (2009a)] as follows.

Lemma 3.13.3. *The map Υ_i is injective if the following conditions are satisfied:*

- *For every edge e of the i-th loop, there is another loop having only e in common with the i-th loop, and*
- *The i-th loop has at most one edge not in common with any other loop.*

Proof. In this situation, all but at most one edge variable appear by themselves as an entry of the i-th row, and the possible last remaining variable appears summed together with the other variables. More explicitly, if $t_{i_1}, \ldots, t_{i_v}$ are the variables corresponding to the edges of a loop ℓ_i, up to rearranging the entries in the corresponding row of $\eta^\dagger \Lambda \eta$ and neglecting other entries, the map Υ_i is given by

$$(t_{i_1}, \ldots, t_{i_v}) \mapsto (t_{i_1} + \cdots + t_{i_v}, \pm t_{i_1}, \ldots, \pm t_{i_v})$$

if ℓ_i has no edge not in common with any other loop, and

$$(t_{i_1}, \ldots, t_{i_v}) \mapsto (t_{i_1} + \cdots + t_{i_v}, \pm t_{i_1}, \ldots, \pm t_{i_{v-1}})$$

if ℓ_i has a single edge t_v not in common with any other loop. In either case the map Υ_i is injective, as claimed. $\square$

Every (finite) graph Γ may be embedded in a compact orientable surface of finite genus. The minimum genus of an orientable surface in which Γ may be embedded is the *genus* of Γ. Thus, Γ is planar if and only if it may be embedded in a sphere, if and only if its genus is 0.

Definition 3.13.4. An embedding of a graph Γ in an orientable surface S is a *2-cell embedding* if the complement of Γ in S is homeomorphic to a union of open 2-cells (the *faces*, or *regions* determined by the embedding). An embedding of Γ in S is a *closed 2-cell embedding* if the closure of every face is a disk.

It is known that an embedding of a connected graph is minimal genus if and only if it is a 2-cell embedding ([Mohar and Thomassen (2001)], Proposition 3.4.1 and Theorem 3.2.4). We discuss below conditions on the existence of *closed* 2-cell embeddings, *cf.* [Mohar and Thomassen (2001)], §5.5.

For our purposes, the advantage of having a closed 2-cell embedding for a graph Γ is that the faces of such an embedding determine a choice of loops of Γ, by taking the boundaries of the 2-cells of the embedding together with a basis of generators for the homology of the Riemann surface in which the graph is embedded.

Lemma 3.13.5. *A closed 2-cell embedding $\iota : \Gamma \to S$ of a connected graph Γ on a surface of (minimal) genus g, together with the choice of a face of the embedding and a basis for the homology $H_1(S, \mathbb{Z})$ determine a basis of $H_1(\Gamma, \mathbb{Z})$ given by $2g + f - 1$ loops, where f is the number of faces of the embedding.*

Proof. Orient (arbitrarily) the edges of Γ and the faces, and then add the edges on the boundary of each face with sign determined by the orientations. The fact that the closure of each face is a 2-disk guarantees that the boundary is null-homotopic. This produces a number of loops equal to the number f of faces. It is clear that these f loops are *not* independent: the sum of any $f - 1$ of them must equal the remaining one, up to sign. Any $f - 1$ loops, however, will be independent in $H_1(\Gamma)$. Indeed, these $f - 1$ loops, together with $2g$ generators of the homology of S, generate $H_1(\Gamma)$. The homology group $H_1(\Gamma)$ has rank $2g + f - 1$, as one can see from the Euler characteristic formula

$$b_0(S) - b_1(S) + b_2(S) = 2 - 2g$$

$$= \chi(S) = v - e + f = b_0(\Gamma) - b_1(\Gamma) + f = 1 - \ell + f,$$

so there will be no other relations. $\square$

One refers to the chosen one among the f faces as the "external face" and the remaining $f - 1$ faces as the "internal faces".

Thus, given a closed 2-cell embedding $\iota : \Gamma \to S$, we can use a basis of $H_1(\Gamma, \mathbb{Z})$ constructed as in Lemma 3.13.5 to compute the map Υ and the maps Υ_i of (3.13.1). We then have the following result.

Lemma 3.13.6. *Assume that Γ is closed-2-cell embedded in a surface. With notation as above, assume that*

- *any two of the f faces have at most one edge in common.*

Then the $f - 1$ maps Υ_i, defined with respect to a choice of basis for $H_1(\Gamma)$ as in Lemma 3.13.5, are all injective. If further

- *every edge of Γ is in the boundary of two of the f faces,*

then Υ is injective.

Proof. The injectivity of the $f - 1$ maps Υ_i follows from Lemma 3.13.3. If ℓ is a loop determined by an internal face, the variables corresponding to edges in common between ℓ and any other internal loop will appear as $(\pm)$ individual entries on the row corresponding to ℓ. Since ℓ has at most one edge in common with the external region, this accounts for all but at most one of the edges in ℓ. By Lemma 3.13.3, the injectivity of Υ_i follows. Finally, as shown in Lemma 3.13.2, the map Υ is injective if every edge is in one of the $f - 1$ loops and the $f - 1$ maps Υ_i are injective. The stated condition guarantees that the edge appears in the loops corresponding to the faces separated by that edge. At least one of them is internal, so that every edge is accounted for. □

Recall the notions of connectivity of graphs that we discussed in the first chapter in relation to the 1PI condition, Definition 1.5.1.

We also recall the notion of *face width* for an embedding of a graph in a Riemann surface.

Definition 3.13.7. Let $\iota : \Gamma \hookrightarrow S$ be a given embedding of a graph Γ on a Riemann surface S. The face width $fw(\Gamma, \iota)$ is the largest number $k \in \mathbb{N}$ such that every non-contractible simple closed curve in S intersects Γ at least k times. When S is a sphere, hence $\iota : \Gamma \hookrightarrow S$ is a planar embedding, one sets $fw(\Gamma, \iota) = \infty$.

For a graph Γ with at least 3 vertices and with no looping edges, the condition that an embedding $\iota : \Gamma \hookrightarrow S$ is a *closed* 2-cell embedding is equivalent to the properties that Γ is 2-vertex-connected and that the embedding has face width $fw(\Gamma, \iota) \geq 2$; see Proposition 5.5.11 of [Mohar and Thomassen (2001)]. It is not known whether every 2-vertex-connected graph Γ admits a closed 2-cell embedding. The "strong orientable embedding conjecture" states that this is the case, namely, that every 2-vertex-connected graph Γ admits a closed 2-cell embedding in some orientable surface S, of face width at least two (see [Mohar and Thomassen (2001)], Conjecture 5.5.16).

Theorem 3.13.8. *The following conditions imply the injectivity of* Υ.

(1) Let Γ be a graph with at least 3 vertices and with no looping edges, which is closed-2-cell embedded in an orientable surface S. Then, if any two of the faces have at most one edge in common, the map Υ is injective.

(2) Let Γ be a 3-edge-connected graph, with at least 3 vertices and no looping edges, admitting a closed-2-cell embedding $\iota : \Gamma \hookrightarrow S$ with face width $fw(\Gamma, \iota) \geq 3$. Then the maps Υ_i, Υ are all injective.

Proof. (1) It suffices to show that, under these conditions on the graph Γ, the second condition of Lemma 3.13.6 is automatically satisfied, so that only the first condition remains to be checked. That is, we show that every edge of Γ is in the boundary of two faces. Assume an edge is not in the boundary of two faces. Then that edge must bound the same face on both of its sides. The closure of the face is a cell, by assumption. Let γ be a path from one side of the edge to the other. Since γ splits the cell into two connected components, it follows that removing the edge splits Γ into two connected components, hence Γ is not 2-edge-connected. However, the fact that Γ has at least 3 vertices and no looping edges and it admits a closed 2-cell embedding implies that Γ is 2-vertex-connected, hence in particular it is 1PI by Lemma 1.5.5, and this gives a contradiction.

(2) The previous result shows that the second condition stated in Lemma 3.13.6 is automatically satisfied, so the only thing left to check is that the first condition stated in Lemma 3.13.6 holds. Assume that two faces F_1, F_2 have more than one edge in common. Since F_1, F_2 are (path)-connected, there are paths γ_i in F_i connecting corresponding sides of the edges. With suitable care, it can be arranged that $\gamma_1 \cup \gamma_2$ is a closed path γ meeting Γ in 2 points. Since the embedding has face width ≥ 3, γ must be null-homotopic in the surface, and in particular it splits it into two connected components. Therefore Γ is split into two connected components by removing the two edges, hence Γ cannot be 3-edge-connected. $\square$

A more geometric description of these combinatorial properties in terms of *wheel neighborhoods* at the vertices of the graph is discussed in [Aluffi and Marcolli (2009a)]. Further, it is shown there that the injectivity property of the map Υ extends from the classes of graphs described in the previous theorem to other graphs obtained from these by simple combinatorial operations such as attaching a looping edge at a vertex of a given graph or subdividing edges by adding intermediate valence two vertices.

The combinatorial conditions of Theorem 3.13.8 are fairly natural from a physical viewpoint. In fact, 2-edge-connected is just the usual 1PI condition, while 3-edge-connected or 2PI is the next strengthening of this condition (the 2PI effective action is often considered in quantum field theory), and the face width condition is also the next strengthening of face width 2, which a well known combinatorial conjecture on graphs [Mohar and Thomassen (2001)] expects should simply follow for graphs that are 2-vertex-connected. (The latter condition is a bit more than 1PI: for graphs with at least two vertices and no looping edges it is equivalent to all the splittings of the graph at vertices also being 1PI, as we showed in Lemma 1.5.5.) The condition that the graph has no looping edges is only a technical device for the proof. In fact, it is then easy to show (see [Aluffi and Marcolli (2009a)]) that adding looping edges does not affect the injectivity of the map Υ.

When the map Υ is injective, it is possible to rephrase the computation of the parametric Feynman integral as a period of the complement of the determinant hypersurface $\hat{\mathcal{D}}_\ell \subset \mathbb{A}^{\ell^2}$. Notice also that if the map Υ is injective then one has a well defined map $\mathbb{P}^{n-1} \to \mathbb{P}^{\ell^2-1}$, which is otherwise not everywhere defined. Motivically, when the map $\Upsilon : \mathbb{A}^n \to \mathbb{A}^{\ell^2}$ is injective, the complexity of Feynman integrals of the graph Γ as a period is controlled by the motive $\mathfrak{m}(\mathbb{A}^{\ell^2} \smallsetminus \hat{\mathcal{D}}_\ell, \hat{\Sigma}_\Gamma \smallsetminus (\hat{\mathcal{D}}_\ell \cap \hat{\Sigma}_\Gamma))$, where $\hat{\Sigma}_\Gamma$ is a normal crossings divisor in $\mathbb{A}^{\ell^2}$ such that $\Upsilon(\partial \sigma_n) \subset \hat{\Sigma}_\Gamma$. We recall below how to construct the divisor $\hat{\Sigma}_\Gamma$.

One can in fact rewrite the Feynman integral (as usual up to a divergent Γ-factor) in the form

$$U(\Gamma) = \int_{\Upsilon(\sigma_n)} \frac{\mathcal{P}_\Gamma(x,p)^{-n+D\ell/2}\omega_\Gamma(x)}{\det(x)^{-n+(\ell+1)D/2}},$$

for a polynomial $\mathcal{P}_\Gamma(x,p)$ on $\mathbb{A}^{\ell^2}$ that restricts to $P_\Gamma(t,p)$, and with $\omega_\Gamma(x)$ the image of the volume form. Let then $\hat{\Sigma}_\Gamma$ be a normal crossings divisor in $\mathbb{A}^{\ell^2}$, which contains the boundary of the domain of integration, $\Upsilon(\partial \sigma_n) \subset \hat{\Sigma}_\Gamma$. The question on the motivic nature of the resulting period can then be reformulated (again modulo divergences) in this case as the question of whether the motive

$$\mathfrak{m}(\mathbb{A}^{\ell^2} \smallsetminus \hat{\mathcal{D}}_\ell, \hat{\Sigma}_\Gamma \smallsetminus (\hat{\Sigma}_\Gamma \cap \hat{\mathcal{D}}_\ell)) \qquad (3.97)$$

is mixed Tate.

The advantage of moving the period computation via the map $\Upsilon = \Upsilon_\Gamma$ from the hypersurface complement $\mathbb{A}^n \smallsetminus \hat{X}_\Gamma$ to the complement of the determinant hypersurface $\mathbb{A}^{\ell^2} \smallsetminus \hat{\mathcal{D}}_\ell$ is that, unlike what happens with the graph hypersurfaces, it is well known that the determinant hypersurface $\hat{\mathcal{D}}_\ell$ is a mixed Tate motive, as we have already seen in detail in Theorems 2.6.2 and 2.6.3 above.

One then sees that, in this approach, the difficulty has been moved from understanding the motivic nature of the hypersurface complement to having some control on the other term of the relative cohomology, namely the normal crossings divisor $\hat{\Sigma}_\Gamma$ and the way it intersects the determinant hypersurface.

If the motive of $\hat{\Sigma}_\Gamma \smallsetminus (\hat{\Sigma}_\Gamma \cap \hat{\mathcal{D}}_\ell)$ is mixed Tate, then knowing that $\mathbb{A}^{\ell^2} \smallsetminus \hat{\mathcal{D}}_\ell$ is always mixed Tate, the fact that mixed Tate motives form a triangulated subcategory of the triangulated category of mixed motives would show that the motive (3.97) whose realization is the relative cohomology would also be mixed Tate.

Thus, Question 3.12.2 can be reformulated, under the combinatorial conditions given above for the embedding, as a question on when the motive of $\hat{\Sigma}_\Gamma \smallsetminus (\hat{\Sigma}_\Gamma \cap \hat{\mathcal{D}}_\ell)$ is mixed Tate. This requires, first of all, a better description of the divisor $\hat{\Sigma}_\Gamma$ that contains the boundary $\partial(\Upsilon(\sigma_n))$.

A first observation in [Aluffi and Marcolli (2009a)] is that one can use the same normal crossings divisor $\hat{\Sigma}_{\ell,g}$ for all graphs Γ with a fixed number of loops and a fixed genus (that is, the minimal genus of an orientable surface in which the graph can be embedded). This divisor is given by a union of linear spaces.

Proposition 3.13.9. *There exists a normal crossings divisor $\hat{\Sigma}_{\ell,g} \subset \mathbb{A}^{\ell^2}$, which is a union of $N = \binom{f}{2}$ linear spaces,*

$$\hat{\Sigma}_{\ell,g} := L_1 \cup \cdots \cup L_N, \tag{3.98}$$

such that, for all graphs Γ with ℓ loops and genus g closed 2-cell embedding, the preimage under $\Upsilon = \Upsilon_\Gamma$ of the union $\hat{\Sigma}_\Gamma$ of a subset of components of $\hat{\Sigma}_{\ell,g}$ is the algebraic simplex Σ_n in $\mathbb{A}^n$. More explicitly, the components of the divisor $\hat{\Sigma}_{\ell,g}$ can be described by the $N = \binom{f}{2}$ equations

$$\begin{cases} x_{ij} = 0 & 1 \leq i < j \leq f-1 \\ x_{i1} + \cdots + x_{i,f-1} = 0 & 1 \leq i \leq f-1, \end{cases} \tag{3.99}$$

where $f = \ell - 2g + 1$ is the number of faces of the embedding of the graph Γ on a surface of genus g.

Proof. The polynomial $\det M_\Gamma(t)$ does not depend on the choice of orientation for the loops of Γ. Thus, we can make the following convenient choice of these orientations. We have chosen a closed 2-cell embedding of Γ into an orientable surface of genus g. Such an embedding has f faces, with $\ell = 2g + f - 1$. We can arrange $M_\Gamma(t)$ so that the first $f - 1$ rows correspond to the $f - 1$ loops determined by the 'internal' faces of the embedding. On each face, we choose the positive orientation (counterclockwise with respect to an outgoing normal vector). Then each edge-variable in common between two faces i, j will appear with a minus sign in the entries (i, j) and (j, i) of $M_\Gamma(t)$. These entries are both in the $(\ell - 2g) \times (\ell - 2g)$ upper-left minor.

We have in fact seen that the injectivity of an $(\ell - 2g) \times (\ell - 2g)$ minor of the matrix M_Γ suffices to control the injectivity of the map Υ, and we can arrange so that the minor is the upper-left part of the $\ell \times \ell$ ambient matrix. Then the hyperplanes in $\mathbb{A}^n$ associated to the coordinates t_i can be obtained by pulling back linear spaces along this minor. On the diagonal of the $(f - 1) \times (f - 1)$ submatrix we find all edges making up each face, with a positive sign. It follows that the pull-backs of the equations (3.99) produce a list of all the edge variables, possibly with redundancies. The components of $\hat{\Sigma}_{\ell,g}$ that form the divisor $\hat{\Sigma}_\Gamma$ are selected by eliminating those components of $\hat{\Sigma}_{\ell,g}$ that contain the image of the graph hypersurface (*i.e.* coming from the zero entries of the matrix $M_\Gamma(t)$). □

A second observation of [Aluffi and Marcolli (2009a)] is then that, using inclusion-exclusion, it suffices to show that arbitrary intersections of the components L_i of $\hat{\Sigma}_{\ell,g}$ have the property that

$$(\cap_{i \in I} L_i) \smallsetminus \hat{\mathcal{D}}_\ell \tag{3.100}$$

is mixed Tate. In fact, this would then imply that also the locus

$$\hat{\Sigma}_\Gamma \smallsetminus (\hat{\mathcal{D}}_\ell \cap \hat{\Sigma}_\Gamma)$$

is mixed Tate, by repeatedly using inclusion–exclusion.

Notice that the intersection $\cap_{i \in I} L_i$ is a linear subspace of codimension $\#I$ in $\mathbb{A}^{\ell^2}$. In general, the intersection of a linear subspace with the determinant is *not* mixed Tate. For example, the intersection of a general $\mathbb{A}^3$ with $\hat{\mathcal{D}}_3$ is a cone over a genus-1 curve. In fact, working projectively, $\mathcal{D}_3$ is a degree-3 hypersurface in $\mathbb{P}^8$, with singularities in codimension > 1. Therefore, the intersection with a general $\mathbb{P}^2$ is a nonsingular cubic curve, therefore a curve of genus $= 1$. The affine version is a cone over this. Thus, in order to understand under what conditions the locus (3.100) will be mixed Tate, we have to understand in what sense the intersections $\cap_{i \in I} L_i$ are special.

The following characterization, which is given in [Aluffi and Marcolli (2009a)], leads to a reformulation of the problem in terms of certain "manifolds of frames".

Lemma 3.13.10. *Let E be a fixed ℓ-dimensional vector space. Every $I \subseteq \{1, \dots, N\}$ determines a choice of linear subspaces $V_1, \dots, V_\ell$ of E with the property that*

$$\cap_{k \in I} L_k = \{(v_1, \dots, v_\ell) \in \mathbb{A}^{\ell^2} \mid \forall i, v_i \in V_i\}. \tag{3.101}$$

Here, we denote an $\ell \times \ell$ matrix in $\mathbb{A}^{\ell^2}$ by its ℓ row-vectors $v_i \in E$.

Further, $\dim V_i \geq i - 1$. Further still, there exists a basis $(e_1, \ldots, e_\ell)$ of E such that each space V_i is the span of a subset (of cardinality $\geq i - 1$) of the vectors e_j.

Proof. Recall (Proposition 3.13.9) that the components L_k of $\hat{\Sigma}_{\ell,g}$ consist of matrices for which either the (i,j) entry x_{ij} equals 0, for $1 \leq i < j \leq \ell - 2g$, or

$$x_{i1} + \cdots + x_{i,\ell-2g} = 0$$

for $1 \leq i \leq \ell - 2g$. Thus, each L_k consists of ℓ-tuples $(v_1, \ldots, v_\ell)$ for which exactly one row v_i belongs to a fixed hyperplane of E, and more precisely to one of the hyperplanes

$$x_1 + \cdots + x_{\ell-2g} = 0, \quad x_2 = 0, \quad \ldots \quad x_{\ell-2g} = 0. \tag{3.102}$$

The statement follows by choosing V_i to be the intersection of the hyperplanes corresponding to the L_k in the i-th row, among those listed in (3.102). Since there are at most $\ell - 2g - i + 1$ hyperplanes L_k in the i-th row,

$$\dim V_i \geq \ell - (\ell - 2g - i + 1) = 2g + i - 1 \geq i - 1 \quad .$$

Finally, to obtain the basis $(e_1, \ldots, e_\ell)$ mentioned in the statement, simply choose the basis dual to the basis $(x_1 + \cdots + x_{\ell-2g}, x_2, \ldots, x_\ell)$ of the dual space to E. $\square$

A sufficient condition that would ensure that the locus $\hat{\Sigma}_\Gamma \smallsetminus (\hat{\Sigma}_\Gamma \cap \hat{\mathcal{D}}_\ell)$ is mixed Tate is then described in terms of *manifolds of frames*.

Definition 3.13.11. For a given ambient space $V = \mathbb{A}^\ell$ and an assigned collection of linear subspaces V_i, $i = 1, \ldots, \ell$, the manifold of frames $\mathbb{F}(V_1, \ldots, V_\ell)$ is defined as the locus

$$\mathbb{F}(V_1, \ldots, V_\ell) := \{(v_1, \ldots, v_\ell) \in \mathbb{A}^{\ell^2} \mid v_k \in V_k\} \cap (\mathbb{A}^{\ell^2} \smallsetminus \hat{\mathcal{D}}_\ell). \tag{3.103}$$

More generally, $\mathbb{F}(V_1, \ldots, V_r) \subset V_1 \times \cdots \times V_r$ denotes the locus of r-tuples of linearly independent vectors in a given vector space V, where each v_i is constrained to belong to the given subspace V_i.

The previous characterization of the locus (3.100), together with the use of inclusion–exclusion arguments, shows that if, for a fixed ℓ, the manifold of frames is mixed Tate for all choices of the subspaces V_i, this suffices to guarantee that the motive

$$\mathfrak{m}(\mathbb{A}^{\ell^2} \smallsetminus \hat{\mathcal{D}}_\ell, \hat{\Sigma}_{\ell,g} \smallsetminus (\hat{\Sigma}_{\ell,g} \cap \hat{\mathcal{D}}_\ell)) \tag{3.104}$$

is also mixed Tate. When this is the case, the result then implies that, modulo divergences, the residues of Feynman graphs Γ with ℓ loops and genus g, for which the injectivity condition on Υ holds, are all periods of mixed Tate motives.

In the case of two and three loops, one can verify explicitly that the manifolds of frames $\mathbb{F}(V_1, V_2)$ and $\mathbb{F}(V_1, V_2, V_3)$ are mixed Tate. This is done in [Aluffi and Marcolli (2009a)] by exhibiting an explicit stratification, from which, as in the case of the determinant hypersurface discussed above, one can show that the motive defines an object in the category of mixed Tate motives as a subtriangulated category of the category of mixed motives. One can also compute explicitly the corresponding classes in the Grothendieck ring, as a function of the Lefschetz motive $\mathbb{L}$. They are of the following form.

Proposition 3.13.12. *For given subspaces V_1, V_2, the manifold of frames $\mathbb{F}(V_1, V_2)$ is a mixed Tate motive, whose class in the Grothendieck ring is*

$$[\mathbb{F}(V_1, V_2)] = \mathbb{L}^{d_1 + d_2} - \mathbb{L}^{d_1} - \mathbb{L}^{d_2} - \mathbb{L}^{d_{12}+1} + \mathbb{L}^{d_{12}} + \mathbb{L},$$

with $d_i = \dim(V_i)$ and $d_{ij} = \dim(V_i \cap V_j)$.

Proof. We want to parameterize all pairs (v_1, v_2) of vectors such that $v_1 \in V_1$, $v_2 \in V_2$, and $\dim\langle v_1, v_2 \rangle = 2$. This locus can be decomposed into the two (possibly empty) pieces

(1) $v_1 \in V_1 \smallsetminus (V_1 \cap V_2)$, and $v_2 \in V_2 \smallsetminus \{0\}$;
(2) $v_1 \in (V_1 \cap V_2) \smallsetminus \{0\}$, and $v_2 \in V_2 \smallsetminus \langle v_1 \rangle$.

This exhausts all the possible ways of obtaining linearly independent vectors with the first one in V_1 and the second one in V_2 and the manifold of frames $\mathbb{F}(V_1, V_2)$ is the union of these two loci. The first case describes the locus

$$(V_1 \smallsetminus (V_1 \cap V_2)) \times (V_2 \smallsetminus \{0\})$$

which is clearly mixed Tate, being obtained by taking products and complements of affine spaces. Its class in the Grothendieck ring is of the form

$$(\mathbb{L}^{d_1} - \mathbb{L}^{d_{12}})(\mathbb{L}^{d_2} - 1).$$

The locus defined by the second case can be described equivalently by the following procedure. Consider the projective space $\mathbb{P}(V_1 \cap V_2)$, and the trivial bundles $\mathcal{V}_{12} \subseteq \mathcal{V}_2$ with fiber $V_1 \cap V_2 \subseteq V_2$. The tautological line bundle $\mathcal{O}_{12}(-1)$ over $\mathbb{P}(V_1 \cap V_2)$ sits inside $\mathcal{V}_{12}$, hence inside $\mathcal{V}_2$. The desired pairs of vectors (v_1, v_2) are obtained by choosing a point $p \in \mathbb{P}(V_1 \cap V_2)$, a vector $v_1 \neq 0$ in the fiber of $\mathcal{O}_{12}(-1)$ over p, and a vector v_2 in the fiber of $V_2 \smallsetminus \mathcal{O}_{12}(-1)$ over p. This again defines a mixed Tate locus, using the same two properties of the existence of distinguished triangles in the category of mixed motives associated to closed embeddings and homotopy invariance, as in Theorems 2.6.2 and 2.6.3. The homotopy invariance property, which is formulated for products, extends in fact to the locally trivial case of vector bundles or projective bundles, see [Voevodsky (2000)]. The class in the Grothendieck ring, in this case, is then given by the expression

$$(\mathbb{L}^{d_{12}} - 1)(\mathbb{L}^{d_2} - \mathbb{L}).$$

Thus, the class of $\mathbb{F}(V_1, V_2)$ is the sum of these two classes,

$$[\mathbb{F}(V_1, V_2)] = (\mathbb{L}^{d_1 + d_2} - \mathbb{L}^{d_1} - \mathbb{L}^{d_2 + d_{12}} + \mathbb{L}^{d_{12}}) + (\mathbb{L}^{d_2 + d_{12}} - \mathbb{L}^{d_{12} + 1} - \mathbb{L}^{d_2} + \mathbb{L})$$

$$= \mathbb{L}^{d_1 + d_2} - \mathbb{L}^{d_1} - \mathbb{L}^{d_2} - \mathbb{L}^{d_{12} + 1} + \mathbb{L}^{d_{12}} + \mathbb{L}. \qquad \square$$

The case of three subspaces already requires a more delicate analysis of the contribution of the various strata, as in §6.2 of [Aluffi and Marcolli (2009a)], which we reproduce here below.

Theorem 3.13.13. *Let V_1, V_2, V_3 be assigned subspaces of a vector space V. Then the manifold of frames $\mathbb{F}(V_1, V_2, V_3)$ is a mixed Tate motive, with class in the Grothendieck ring given by*

$$[\mathbb{F}(V_1, V_2, V_3)] = (\mathbb{L}^{d_1} - 1)(\mathbb{L}^{d_2} - 1)(\mathbb{L}^{d_3} - 1)$$

$$-(\mathbb{L} - 1)((\mathbb{L}^{d_1} - \mathbb{L})(\mathbb{L}^{d_{23}} - 1) + (\mathbb{L}^{d_2} - \mathbb{L})(\mathbb{L}^{d_{13}} - 1) + (\mathbb{L}^{d_3} - \mathbb{L})(\mathbb{L}^{d_{12}} - 1)$$

$$+(\mathbb{L} - 1)^2(\mathbb{L}^{d_1 + d_2 + d_3 - D} - \mathbb{L}^{d_{123} + 1}) + (\mathbb{L} - 1)^3,$$

where $d_i = \dim(V_1)$, $d_{ij} = \dim(V_i \cap V_j)$, $d_{ijk} = \dim(V_i \cap V_j \cap V_k)$, and $D = D_{ijk} = \dim(V_i + V_j + V_k)$.

Proof. This time one can proceed in the following way to obtain a stratification. One can look for a stratification $\{S_\alpha\}$ of V_3 which is finer than the one induced by the subspace arrangement $V_1 \cap V_3, V_2 \cap V_3$ and such that, for v_3 in S_α, the class $\mathbb{F}_\alpha := [\mathbb{F}(\pi(V_1), \pi(V_2))]$, with $\pi : V \to V' := V/\langle v_3 \rangle$ the projection, depends only on α and not on the chosen vector $v_3 \in S_\alpha$. In other words, we want to construct a stratification of V_3 such that the dimensions of the spaces $\pi(V_1)$, $\pi(V_2)$ and $\pi(V_1 \cap V_2)$ are constant along the strata. The following five loci define such a stratification of $V_3 \smallsetminus \{0\}$:

(1) $S_{123} := (V_1 \cap V_2 \cap V_3) \smallsetminus \{0\}$;

(2) $S_{13} := (V_1 \cap V_3) \smallsetminus (V_1 \cap V_2 \cap V_3);$

(3) $S_{23} := (V_2 \cap V_3) \smallsetminus (V_1 \cap V_2 \cap V_3);$

(4) $S_{(12)3} := ((V_1 + V_2) \cap V_3) \smallsetminus ((V_1 \cup V_2) \cap V_3);$

(5) $S_3 := V_3 \smallsetminus ((V_1 + V_2) \cap V_3).$

The dimensions are indeed constant along the strata and given by

	$\dim \pi(V_1)$	$\dim \pi(V_2)$	$\dim(\pi(V_1) \cap \pi(V_2))$
S_{123}	$d_1 - 1$	$d_2 - 1$	$d_{12} - 1$
S_{13}	$d_1 - 1$	d_2	d_{12}
S_{23}	d_1	$d_2 - 1$	d_{12}
$S_{(12)3}$	d_1	d_2	$d_{12} + 1$
S_3	d_1	d_2	d_{12}

This information can then be translated into explicit classes $[\mathbb{F}_\alpha]$ in the Grothendieck ring associated to the strata S_α so that the class of the frame manifold is of the form

$$[\mathbb{F}(V_1, V_2, V_3)] = \sum_\alpha \mathbb{L}^{s_\alpha} [\mathbb{F}_\alpha][S_\alpha], \qquad (3.105)$$

where the number s_α is the number of subspaces V_i, $i = 1, 2$ containing the vector $v_3 \in S_\alpha$. This is also independent of the choice of v_3 and only dependent on α by the properties of the stratification. For the table of dimensions obtained above the corresponding classes $[\mathbb{F}_\alpha]$ are as follows.

	$[\mathbb{F}_\alpha]$
S_{123}	$\mathbb{L}^{d_1+d_2-2} - \mathbb{L}^{d_1-1} - \mathbb{L}^{d_2-1} - \mathbb{L}^{d_{12}} + \mathbb{L}^{d_{12}-1} + \mathbb{L}$
S_{13}	$\mathbb{L}^{d_1+d_2-1} - \mathbb{L}^{d_1-1} - \mathbb{L}^{d_2} - \mathbb{L}^{d_{12}+1} + \mathbb{L}^{d_{12}} + \mathbb{L}$
S_{23}	$\mathbb{L}^{d_1+d_2-1} - \mathbb{L}^{d_1} - \mathbb{L}^{d_2-1} - \mathbb{L}^{d_{12}+1} + \mathbb{L}^{d_{12}} + \mathbb{L}$
$S_{(12)3}$	$\mathbb{L}^{d_1+d_2} - \mathbb{L}^{d_1} - \mathbb{L}^{d_2} - \mathbb{L}^{d_{12}+2} + \mathbb{L}^{d_{12}+1} + \mathbb{L}$
S_3	$\mathbb{L}^{d_1+d_2} - \mathbb{L}^{d_1} - \mathbb{L}^{d_2} - \mathbb{L}^{d_{12}+1} + \mathbb{L}^{d_{12}} + \mathbb{L}$

One then obtains the following values for the class of the stratum $[S_\alpha]$ and the number s_α.

	$[S_\alpha]$	s_α
S_{123}	$\mathbb{L}^{d_{123}} - 1$	2
S_{13}	$\mathbb{L}^{d_{13}} - \mathbb{L}^{d_{123}}$	1
S_{23}	$\mathbb{L}^{d_{23}} - \mathbb{L}^{d_{123}}$	1
$S_{(12)3}$	$\mathbb{L}^{d_1+d_2+d_3-D-d_{12}} - \mathbb{L}^{d_{13}} - \mathbb{L}^{d_{23}} + \mathbb{L}^{d_{123}}$	0
S_3	$\mathbb{L}^{d_3} - \mathbb{L}^{d_1+d_2+d_3-D-d_{12}}$	0

Using the formula (3.105) then gives the stated result. Although the stratification is used there only to compute the class in the Grothendieck group, an argument similar to the one of Theorems 2.6.2 and 2.6.3 can be used to show that the same stratifications used to compute $[\mathbb{F}(V_1, V_2)]$ and $[\mathbb{F}(V_1, V_2, V_3)]$ also show that $\mathbb{F}(V_1, V_2)$ and $\mathbb{F}(V_1, V_2, V_3)$ define objects in the triangulated category of mixed Tate motives.   □

One can use these explicit computations of classes of manifolds of frames for two and three subspaces to obtain specific information about the motives
$$\mathfrak{m}(\mathbb{A}^{\ell^2} \smallsetminus \hat{\mathcal{D}}_\ell, \hat{\Sigma}_\Gamma \smallsetminus (\hat{\Sigma}_\Gamma \cap \hat{\mathcal{D}}_\ell))$$
for specific Feynman graphs with up to three loops. For example, we report here briefly the case of the wheel-with-three-spokes graph (the 1-skeleton of a tetrahedron) described in [Aluffi and Marcolli (2009a)].

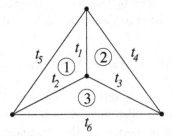

This graph has matrix $M_\Gamma(t)$ given by
$$\begin{pmatrix} t_1 + t_2 + t_5 & -t_1 & -t_2 \\ -t_1 & t_1 + t_3 + t_4 & -t_3 \\ -t_2 & -t_3 & t_2 + t_3 + t_6 \end{pmatrix}.$$
Labeling entries of the matrix as x_{ij}, we can obtain $t_1, \ldots, t_6$ as pull-backs of the following:
$$\begin{cases} t_1 = -x_{12} \\ t_2 = -x_{13} \\ t_3 = -x_{23} \\ t_4 = x_{21} + x_{22} + x_{23} \\ t_5 = x_{11} + x_{12} + x_{13} \\ t_6 = x_{31} + x_{32} + x_{33} \end{cases}$$
Thus, we can consider as $\hat{\Sigma}_\Gamma$ the normal crossings divisor defined by the equation
$$x_{12}x_{13}x_{23}(x_{11} + x_{12} + x_{13})(x_{21} + x_{22} + x_{23})(x_{31} + x_{32} + x_{33}) = 0.$$

In order to compute the class in the Grothendieck ring of the intersection $\hat{\Sigma}_\Gamma \cap (\mathbb{A}^9 \smallsetminus \hat{\mathcal{D}}_3)$, one can use inclusion-exclusion and compute the classes for all the intersections of subsets of components of the divisor $\hat{\Sigma}_\Gamma$. This divisor has 6 components, hence there are $2^6 = 64$ such intersections. Each of them determines a choice of three subspaces V_1, V_2, V_3 corresponding to the linearly independent vectors given by the rows of the matrix (x_{ij}) in $\mathbb{A}^9$. All of the corresponding classes $[\mathbb{F}(V_1, V_2, V_3)]$, for each of the 64 possibilities, were computed explicitly in [Aluffi and Marcolli (2009a)]. We do not reproduce them all here, but we only give the final result, which shows that the resulting class is then of the form

$$[\hat{\Sigma}_\Gamma \smallsetminus (\hat{\mathcal{D}}_3 \cap \hat{\Sigma}_\Gamma)] = \mathbb{L}(6\mathbb{L}^4 - 3\mathbb{L}^3 + 2\mathbb{L}^2 + 2\mathbb{L} - 1)(\mathbb{L} - 1)^3.$$

By arguing as in Theorem 2.6.2, and using the same stratification of the complement $\mathbb{A}^9 \smallsetminus \hat{\mathcal{D}}_3$ induced by the divisor $\hat{\Sigma}_\Gamma$ as in the computation of the class above, one can improve the result from a statement about the class in the Grothendieck ring being a polynomial function of the Lefschetz motive $\mathbb{L}$ to one about the motive $\mathfrak{m}(\hat{\Sigma}_\Gamma \smallsetminus (\hat{\mathcal{D}}_3 \cap \hat{\Sigma}_\Gamma))$ itself being an object in the triangulated subcategory $\mathcal{DMT}_\mathbb{Q}$ of mixed Tate motives inside the triangulated category $\mathcal{DM}_\mathbb{Q}$ of mixed motives. Then this fact, together with the fact that the hypersurface complement itself is a mixed Tate motive, and the distinguished triangle corresponding to the long exact cohomology sequence, suffice to show that the mixed motive

$$\mathfrak{m}(\mathbb{A}^9 \smallsetminus \hat{\mathcal{D}}_3, \hat{\Sigma}_\Gamma \smallsetminus (\hat{\Sigma}_\Gamma \cap \hat{\mathcal{D}}_3))$$

is mixed Tate for the case of the wheel with three spokes. The result for other graphs with three loops can be derived from the same analysis used for the wheel with three spokes, by restricting only to certain components of the divisor, as explained in [Aluffi and Marcolli (2009a)].

For the cases of manifolds of frames $\mathbb{F}(V_1, \ldots, V_r)$ with more than three subspaces, it would seem at first that one should be able to establish an inductive argument that would take care of the cases of more subspaces, but the combinatorics of the possible subspace arrangements quickly becomes very difficult to control. In fact, in general it seems unlikely that the frame manifolds $\mathbb{F}(V_1, \ldots, V_\ell)$ will continue to be mixed Tate for large ℓ and for arbitrarily complex subspace arrangements. The situation may in fact be somewhat similar to that of "Murphy's law in algebraic geometry" described in [Vakil (2006)], where a sufficiently general case may be as bad as possible, while specific cases one can explicitly construct are much better behaved. The connection can probably be made more precise, given that both the result of [Belkale and Brosnan (2003a)] on the general motivic properties of the graph hypersurfaces X_Γ and Murphy's law result of [Vakil (2006)] are based on the same universality result for matroid representations of [Mnëv (1988)]. Another observation from the point of view of matroids is that, while the result of [Belkale and Brosnan (2003a)] is non-constructive, in the sense that it shows that the graph hypersurfaces generate the Grothendieck ring of motives but it does not provide an explicit construction of matroids on which to test the mixed Tate property, it may be that the loci $\hat{\Sigma}_\Gamma \cap (\mathbb{A}^{\ell^2} \setminus \hat{\mathcal{D}}_\ell)$ considered here may provide a possible way to make that general result more explicit, by constructing explicit matroids, along the lines of [Gelfand and Serganova (1987)]. A reformulation of the problem of understanding when frame manifolds are mixed Tate is given in [Aluffi and Marcolli (2009a)] in terms of intersections of unions of Schubert cells in flag varieties. This version of the problem suggests a possible connection to Kazhdan–Lusztig theory [Kazhdan and Lusztig (1980)].

3.14 Handling divergences

All the considerations above on the motivic nature of the relative cohomology involved in the computation of the parametric Feynman integral, either in the hypersurface complement $\mathbb{P}^{n-1} \setminus X_\Gamma$ or in the complement of the determinant hypersurface $\mathbb{P}^{\ell^2-1} \setminus \mathcal{D}_\ell$, assume that the integral is convergent and therefore directly defines a period. In other words, the issue of divergences is not dealt with in this approach. However, one knows very well that most Feynman integrals are divergent, even when in the parametric form one removes the divergent Gamma factor and only looks at the residue, and that some regularization and renormalization procedure is needed to handle these divergences.

In terms of the geometry, after removing the divergent Gamma factor, the source of other possible divergences in the parametric Feynman integral is the locus of intersection of the graph hypersurface X_Γ with the domain of integration σ_n.

Notice that the poles of the integrand that fall inside the integration domain σ_n occur necessarily along the boundary $\partial\sigma_n$, since in the interior the graph polynomial Ψ_Γ takes only strictly positive real values.

Thus, one needs to modify the integrals suitably in such a way as to eliminate, by a regularization procedure, the intersections $X_\Gamma \cap \partial\sigma_n$, or (to work in algebro-geometric terms) the intersections $X_\Gamma \cap \Sigma_n$ which contains the former.

There are different possible ways to achieve such a regularization procedure. We mention here three possible approaches. We discuss the third one in more detail in the following chapter.

One method was developed in [Belkale and Brosnan (2003b)] in the logarithmically divergent case where $n = D\ell/2$, that is, when the polynomial $P_\Gamma(t, p)$ is not present and only the denominator $\Psi_\Gamma(t)^{D/2}$ appears in the parametric Feynman integral. As we have already mentioned, using dimensional regularization, one can, in this case, rewrite the Feynman integral in the form of a local Igusa L-function

$$I(s) = \int_\sigma f(t)^s \omega,$$

for $f = \Psi_\Gamma$. They prove that this L-function has a Laurent series expansion where all the coefficients are periods. In this setting, the issue of eliminating divergences becomes similar to the techniques used, for instance, in the context of log canonical thresholds. The result was more recently extended algorithmically to the non-log-divergent case [Bogner and Weinzierl (2007a)], [Bogner and Weinzierl (2007b)].

Another method, used in [Bloch, Esnault, Kreimer (2006)], consists of eliminating the divergences by separating Σ_n and X_Γ performing a series of blow-ups. Similarly, working in the setting of the determinant hypersurface complement $\mathbb{A}^{\ell^2} \smallsetminus \hat{\mathcal{D}}_\ell$ one can perform blowups to separate the locus of intersection of the determinant hypersurface $\hat{\mathcal{D}}_\ell$ with the divisor $\hat{\Sigma}_{\ell,g}$. To make sure that this operation does not change the motivic properties of the resulting period integrals, one needs to ensure that blow-ups along $\hat{\Sigma}_{\ell,g} \cap \hat{\mathcal{D}}_\ell$ maintain the mixed Tate properties if we know that (3.104) is mixed Tate. For blow-ups performed over a smooth locus one has projective bundles, for which one knows that the mixed Tate property is maintained, since one has an explicit formula for the motive of a projective bundle in the Voevodsky category, which is a sum of Tate twisted copies of the motive of the base, but when the locus is singular, as is typically the case here, then the required analysis is more delicate.

Yet another method was proposed in [Marcolli (2008)], based on deformations instead of resolutions. By considering the graph hypersurface X_Γ as the special fiber X_0 of a family X_s of varieties defined by the level sets $f^{-1}(s)$, for $f = \Psi_\Gamma : \mathbb{A}^n \to \mathbb{A}^1$, one can form a tubular neighborhood

$$D_\epsilon(X) = \cup_{s \in \Delta_\epsilon^*} X_s,$$

for Δ_ϵ^* a punctured disk of radius ϵ, and a circle bundle $\pi_\epsilon : \partial D_\epsilon(X) \to X_\epsilon$. One can then regularize the Feynman integral by integrating "around the singularities" in the fiber $\pi_\epsilon^{-1}(\sigma \cap X_\epsilon)$. The regularized integral has a Laurent series expansion in the parameter ϵ.

Motivically, one should take into account the fact that the presence of singularities in the hypersurface X_Γ increases the likelihood that the part of the cohomology involved in the period computation can be mixed Tate. In fact, if these were smooth hypersurfaces, they would typically not be mixed Tate even for a small number of loops, while it is precisely because of the fact that they are highly singular that the X_Γ continue to be mixed Tate for a fairly large number of loops and even though one knows that eventually one will run into graphs for which X_Γ is no longer mixed Tate, the fact that they continue to be singular in low codimension stil makes it possible to have a significant part of the cohomology that will still be a realization of a mixed Tate motive.

The use of deformations X_ϵ to regularize the integral, which we describe more in detail in the following chapter, is natural from the point of view of singularity theory, where one works with Milnor fibers of singularites. The singularities of the hypersurfaces X_Γ are non-isolated, but many techniques of singularity theory are designed to cover also this more general case.

From this viewpoint, it is not clear though how to perform the regularization and subtraction of divergences in a way that does not alter the motivic properties. In fact, while the special singular fiber may be mixed Tate, the generic fiber in a family used as deformation can be a general hypersurface that will not be motivically mixed Tate any longer. However, there are notions such as a motivic tubular neighborhood [Levine (2005)] that may be useful in this context, as well as the theory of limiting mixed Hodge structures for a degeneration of a family of algebraic varieties. It is not yet clear how to use this type of methods to obtain motivic information on Feynman integrals after subtraction of divergences.

In general, as we discuss more at length in Chapter 4.2, a regularization procedure for Feynman integrals replaces a divergent integral with a function of some regularization parameters (such as the complexified dimension of DimReg, or the deformation parameter ϵ in the example here above) in which the resulting function has a Laurent series expansion around the pole that corresponds to the divergent integral originally considered. One then uses a procedure of extraction of finite values to eliminate the polar parts of these Laurent series in a way that is *consistent over graphs*, that is, a renormalization procedure.

3.15 Motivic zeta functions and motivic Feynman rules

As we mentioned briefly in describing the Grothendieck ring as a universal Euler characteristic, the counting of points of a variety over finite fields is an example of an additive invariant that, as such, factors through the Grothendieck ring. The behavior of the number of points over finite fields is one of the properties of a variety that reveal its motivic nature. For example, if a variety, say defined over $\mathbb{Q}$, is motivically mixed Tate, then the number of points of the reductions mod p is polynomial in p.

The information on the number of points over finite fields is conveniently packaged in the form of a zeta function

$$Z_X(t) = \exp\left(\sum_n \frac{\#X(\mathbb{F}_{q^n})}{n} t^n\right) \qquad (3.106)$$

Notice that this can be written equivalently in terms of symmetric products as

$$Z_X(t) = \sum_{n \geq 0} \#s^n(X)(\mathbb{F}_q) t^n,$$

where $s^n(X)$ denotes the n-th symmetric power of X.

The fact that the counting of the number of points behaves like an Euler characteristic, hence it descends from the Grothendieck ring, suggested that the zeta function itself may be lifted at the motivic level. This was done in [Kapranov (2000)], where the *motivic zeta function* is defined as

$$Z_X(t) = \sum_{n \geq 0} [s^n(X)] t^n \qquad (3.107)$$

where $[s^n(X)]$ are the classes in the Grothendieck ring and the zeta function can be seen as an element of $K_0(\mathcal{V}_{\mathbb{K}})[[t]]$. For example, one has

$$Z_{\mathbb{P}^1}(t) = (1-t)^{-1}(1 - \mathbb{L}t)^{-1}.$$

Kapranov proved that, when X is the motive of a curve, then the zeta function is a rational function, in the sense that, given a motivic measure $\mu : K_0(\mathcal{M}) \to \mathcal{R}$, the zeta function $Z_{X,\mu}(T) \in \mathcal{R}[[T]]$ is a rational function of T. Later, it was proved in [Larsen and Lunts (2003)] that in general this is not true in the case of algebraic surfaces.

One recovers the zeta function of the variety, for a finite field $\mathbb{K} = \mathbb{F}_q$, by applying a "counting of points" homomorphism $K_0(\mathcal{V}_{\mathbb{K}}) \to \mathbb{Z}$ to the motivic zeta function. One obtains other kinds of zeta functions by applying other "motivic measures" $\mu : K_0(\mathcal{V}_{\mathbb{K}}) \to \mathcal{R}$, which give

$$Z_{X,\mathcal{R}}(t) = \sum_{n \geq 0} \mu(s^n(X)) t^n \in \mathcal{R}[[t]].$$

It is well known that the motivic zeta function contains all the information on the behavior of the number of points of reductions mod p, hence potentially the information on the mixed Tate nature of a variety. A recent survey of these ideas is given in [André (2009)], where it is also observed that the motivic zeta function should play a useful role in the case of motives associated to Feynman graphs.

There are other reasons why one can expect that, indeed, Kapranov's motivic zeta functions may be especially suitable tools to investigate motivic properties of Feynman integrals. One is, for example, the observation made at the end of [Marcolli (2008)] which suggests that the function Ψ_Γ^s, for s a complex variable, that appears in the dimensional regularization of parametric Feynman integrals, regarded as a zeta function

$$Z_\Gamma(T) := \sum_{n \geq 0} \frac{\log^n \Psi_\Gamma}{n!} T^n = \Psi_\Gamma^T,$$

may be related to a motivic zeta function

$$Z_{\mathrm{Log},\Gamma}(T) := \sum_{n \geq 0} s^n (\mathrm{Log}_\Gamma) T^n,$$

where the motive Log_Γ is the pullback of the logarithmic motive (2.39), via the map Ψ_Γ.

Another reason is the formulation of parametric Feynman integrals as local Igusa L-functions given in [Belkale and Brosnan (2003b)]. There is a motivic Igusa L-function constructed by [Denef and Loeser (1998)], which may provide the right tool for a motivic formulation of the dimensionally regularized parametric Feynman integrals.

In quantum field theory it is customary to consider the full partition function of the theory, arranged in an asymptotic series by loop number, instead of looking only at the contribution of individual Feynman graphs. Besides the loop number $\ell = b_1(\Gamma)$, other suitable gradings $\delta(\Gamma)$ are given by the number $n = \#E_{int}(\Gamma)$ of internal edges, or by $\#E_{int}(\Gamma) - b_1(\Gamma) = \#V(\Gamma) - b_0(\Gamma)$, the number of vertices minus the number of connected components. As shown in [Connes and Marcolli (2008)] p.77, these all define gradings on the Connes–Kreimer Hopf algebra of Feynman graphs.

When one considers motivic Feynman rules, these partition functions appear to be interesting analogs of the motivic zeta functions considered in [Kapranov (2000)], [Larsen and Lunts (2003)]. For instance, one can consider a partition function given by the formal series

$$Z(t) = \sum_{N \geq 0} \sum_{\delta(\Gamma)=N} \frac{U(\Gamma)}{\#\mathrm{Aut}(\Gamma)} \, t^N, \qquad (3.108)$$

where $\delta(\Gamma)$ is any one of the gradings described above and where $U(\Gamma) = [\mathbb{A}^n \smallsetminus \hat{X}_\Gamma] \in K_0(\mathcal{V}_{\mathbb{K}})$. Given a motivic measure $\mu : K_0(\mathcal{V}_{\mathbb{K}}) \to \mathcal{R}$, this gives a zeta function with values in $\mathcal{R}[[t]]$ of the form

$$Z_{\mathcal{R}}(t) = \sum_{N \geq 0} \sum_{\delta(\Gamma)=N} \frac{\mu(U(\Gamma))}{\#\mathrm{Aut}(\Gamma)} \, t^N.$$

In the case of the sums over graphs

$$S_N = \sum_{\#V(\Gamma)=N} [X_\Gamma] \frac{N!}{\#\mathrm{Aut}(\Gamma)}$$

considered in [Bloch (2008)], one finds that cancellations between the classes occur in the Grothendieck ring $K_0(\mathcal{V}_{\mathbb{K}})$ and the resulting class S_N is always in the Tate subring $\mathbb{Z}[\mathbb{L}]$ even if the individual terms $[X_\Gamma]$ may be non-mixed Tate. Similarly, it is possible that interesting cancellations of a similar nature may occur in suitable evaluations of the zeta functions (3.108).

Chapter 4

Feynman integrals and Gelfand–Leray forms

We focus in this chapter, following [Marcolli (2008)], on the similarities between the parametric form of the Feynman integrals and the oscillatory integrals used in singularity theory to describe integrals of holomorphic forms over vanishing cycles of a singularity and to relate these to mixed Hodge structures, see [Arnold, Goryunov, Lyashko, Vasilev (1998)] and Vol.II of [Arnold, Gusein-Zade, Varchenko (1988)]. In particular, we show that dimensionally regularized parametric Feynman integrals can be related to the Mellin transform of a Gelfand–Leray form, whose Fourier transform is an oscillatory integral of the sort considered in the context of singularity theory. We show that one can define in this way a regularization procedure for the parametric Feynman integrals, based on the use of Leray cocycles, aimed at eliminating the divergences coming from the intersections $X_\Gamma \cap \sigma_n$, as mentioned in §3.14 above.

4.1 Oscillatory integrals

An *oscillatory integral* is an expression of the form

$$I(\alpha) = \int_{\mathbb{R}^n} e^{i\alpha f(x)} \phi(x)\, dx_1 \cdots dx_n, \qquad (4.1)$$

where $f : \mathbb{R}^n \to \mathbb{R}$ and $\phi : \mathbb{R}^n \to \mathbb{R}$ are smooth functions and $\alpha \in \mathbb{R}^*_+$ is a real parameter. It is well known that, if the phase $f(x)$ is an analytic function in a neighborhood of a critical point x_0, then (4.1) has an asymptotic development for $\alpha \to \infty$ given by a series

$$I(\alpha) \sim e^{i\alpha f(x_0)} \sum_u \sum_{k=0}^{n-1} a_{k,u}(\phi)\, \alpha^u (\log \alpha)^k, \qquad (4.2)$$

where u runs over a finite set of arithmetic progressions of negative rational numbers depending only on the phase $f(x)$, and the $a_{k,u}$ are distributions supported on the critical points of the phase, *cf.* §2.6.1, Vol.II of [Arnold, Gusein-Zade, Varchenko (1988)].

It is also well known that the integral (4.1) can be reformulated in terms of one-dimensional integrals using the *Gelfand–Leray forms*, defined as follows.

Definition 4.1.1. Let X_t be the level hypersurfaces $X_t = \{x \mid f(x) = t\}$. Then the Gelfand–Leray form $\omega_f(x,t)$ of f is the unique $(n-1)$-form on the level hypersurface X_t with the property that

$$df \wedge \omega_f(x,t) = dx_1 \wedge \cdots \wedge dx_n. \tag{4.3}$$

Notice that there is an ambiguity given by different possible choices of an $(n-1)$-form satisfying (4.3), but their restrictions to X_t all agree so that the Gelfand–Leray form on X_t is well defined.

More generally, given a form α, we say that α admits a Gelfand–Leray form with respect to f if there is a differential form, which we denote by α/df, satisfying

$$df \wedge \frac{\alpha}{df} = \alpha. \tag{4.4}$$

As above, the restriction of the form $(\alpha/df)(x,t)$ to X_t is uniquely defined. The form $\omega_f(x,t)$ above is the Gelfand–Leray form of the volume form $\alpha = dx_1 \wedge \cdots \wedge dx_n$, hence one writes it with the notation

$$\omega_f(x,t) = \frac{dx_1 \wedge \cdots \wedge dx_n}{df}. \tag{4.5}$$

The Gelfand–Leray form is given by the *Poincaré residue*

$$\frac{\alpha}{df} = \mathrm{Res}_{\epsilon=0} \frac{\alpha}{f - \epsilon}. \tag{4.6}$$

The Gelfand–Leray function is the associated function

$$J(t) := \int_{X_t} \phi(x)\omega_f(x,t). \tag{4.7}$$

The oscillatory integral, expressed in terms of Gelfand–Leray forms, is then written as

$$I(\alpha) = \int_{\mathbb{R}} e^{i\alpha t} \left(\int_{X_t(\mathbb{R})} \phi(x)\omega_f(x,t) \right) dt, \tag{4.8}$$

where $X_t(\mathbb{R}) \subset \mathbb{R}^n$ is the level set $X_t(\mathbb{R}) = \{x \in \mathbb{R}^n : f(x) = t\}$ and $\omega_f(x, t)$ is the Gelfand–Leray form of f. For more details, see §2.6 and §2.7, Vol.II [Arnold, Gusein-Zade, Varchenko (1988)].

Notice that, up to throwing away a set of measure zero, we can assume here that the integration is over the values $t \in \mathbb{R}$ such that the level set X_t is a smooth hypersurface.

4.2 Leray regularization of Feynman integrals

We consider here the case of complex hypersurfaces $X \subset \mathbb{A}^n = \mathbb{C}^n$, with defining polynomial equation $f = 0$ (where f will be the graph polynomial Ψ_Γ) and the hypersurface complement $\mathcal{D}(f) \subset \mathbb{A}^n$, such that the restriction of f to the interior of the domain of integration $\sigma_n \subset \mathbb{A}^n$ takes strictly positive real values.

Definition 4.2.1. The Leray coboundary $\mathcal{L}(\sigma)$ of a k-chain σ in X is a $(k+1)$-chain in $\mathcal{D}(f)$ obtained by considering a tubular neighborhood of X in $\mathbb{A}^n$, in the following way. If X is a smooth hypersurface, the boundary of its tubular neighborhood is a circle bundle over X. One considers the preimage of σ under the projection map as a chain in $\mathcal{D}(f)$.

The Leray coboundary $\mathcal{L}(\sigma)$ is a cycle if σ is a cycle, and if one changes σ by a boundary then $\mathcal{L}(\sigma)$ also changes by a boundary.

Lemma 4.2.2. *Let σ_ϵ be a k-chain in $X_\epsilon = \{t \in \mathbb{A}^n | f(t) = \epsilon\}$ and let $\mathcal{L}(\sigma_\epsilon)$ be its Leray coboundary in $\mathcal{D}(f - \epsilon)$. Then, for a form $\alpha \in \Omega^k$ that admits a Gelfand–Leray form, one has*

$$\frac{1}{2\pi i} \int_{\mathcal{L}(\sigma(\epsilon))} df \wedge \frac{\alpha}{f - \epsilon} = \int_{\sigma(\epsilon)} \alpha, \tag{4.9}$$

where

$$\frac{d}{d\epsilon} \int_{\sigma(\epsilon)} \alpha = \int_{\sigma(\epsilon)} \frac{d\alpha}{df} - \int_{\partial\sigma(\epsilon)} \frac{\alpha}{df}. \tag{4.10}$$

Proof. First let us show that if α has a Gelfand–Leray form then $d\alpha$ also does. We have a form α/df such that

$$df \wedge \frac{\alpha}{df} = \alpha.$$

Its differential gives

$$d\alpha = d\left(df \wedge \frac{\alpha}{df}\right) = -df \wedge d\left(\frac{\alpha}{df}\right).$$

Thus, the form

$$\frac{d\alpha}{df} = -d\left(\frac{\alpha}{df}\right)$$

is a Gelfand–Leray form for $d\alpha$.

Then we proceed to prove the first statement. One can write

$$\frac{1}{2\pi i} \int_{\mathcal{L}(\sigma(\epsilon))} df \wedge \frac{\alpha}{f - \epsilon} = \frac{1}{2\pi i} \int_{\gamma} \left(\int_{\sigma(s)} \alpha \right) \frac{ds}{s - \epsilon},$$

where $\gamma \cong S^1$ is the boundary of a small disk centered at $\epsilon \in \mathbb{C}$. This can then be written as

$$= \frac{1}{2\pi i} \int_{\gamma} \int_{\sigma(\epsilon)} \alpha \frac{ds}{s - \epsilon} + \left(\frac{1}{2\pi i} \int_{\gamma} \int_{\sigma(s)} \alpha \frac{ds}{s - \epsilon} - \frac{1}{2\pi i} \int_{\gamma} \int_{\sigma(\epsilon)} \alpha \frac{ds}{s - \epsilon} \right).$$

The last term can be made arbitrarily small, so one gets (4.9). To obtain (4.10) notice that

$$\frac{1}{2\pi i} \frac{d}{d\epsilon} \int_{\mathcal{L}(\sigma(\epsilon))} df \wedge \frac{\alpha}{f - \epsilon} = \frac{1}{2\pi i} \int_{\mathcal{L}(\sigma(\epsilon))} df \wedge \frac{\alpha}{(f - \epsilon)^2}.$$

One then uses

$$d\left(\frac{\alpha}{f - \epsilon}\right) = \frac{d\alpha}{f - \epsilon} - \frac{\alpha}{(f - \epsilon)^2}$$

to rewrite the above as

$$\frac{1}{2\pi i} \left(\int_{\mathcal{L}(\sigma(\epsilon))} \frac{d\alpha}{f - \epsilon} - \int_{\mathcal{L}(\sigma(\epsilon))} d\left(\frac{\alpha}{f - \epsilon}\right) \right)$$

$$= \frac{1}{2\pi i} \int_{\mathcal{L}(\sigma(\epsilon))} \frac{df \wedge \frac{d\alpha}{df}}{f - \epsilon} - \frac{1}{2\pi i} \int_{\mathcal{L}(\partial\sigma(\epsilon))} \frac{\alpha}{f - \epsilon}$$

$$= \frac{1}{2\pi i} \int_{\mathcal{L}(\sigma(\epsilon))} \frac{df \wedge \frac{d\alpha}{df}}{f - \epsilon} - \frac{1}{2\pi i} \int_{\mathcal{L}(\partial\sigma(\epsilon))} \frac{df \wedge \frac{\alpha}{df}}{f - \epsilon},$$

where $d\alpha/df$ is a Gelfand–Leray form such that

$$df \wedge \frac{d\alpha}{df} = d\alpha,$$

and α/df is a Gelfand–Leray form with the property that

$$df \wedge \frac{\alpha}{df} = \alpha.$$

This then gives by (4.9)

$$\frac{d}{d\epsilon} \int_{\sigma(\epsilon)} \alpha = \int_{\sigma(\epsilon)} \frac{d\alpha}{df} - \int_{\partial\sigma(\epsilon)} \frac{\alpha}{df}.$$

This completes the proof. $\square$

The formulation given in Proposition 3.4.3 of the parametric Feynman integrals suggests a regularization procedure based on Leray coboundaries, which has the effect of replacing a divergent integral with a meromorphic function of a regularization parameter ϵ, similarly to what happens in the case of DimReg. We know that the divergences of the parametric Feynman integral appear at the intersections $X_\Gamma \cap \sigma_n$, so we can concentrate only on the part of the integral that is supported near this locus.

As mentioned briefly in §3.14 above, one considers a neighborhood $D_\epsilon(X)$ of a hypersurface X in a projective space $\mathbb{P}^{n-1}$, given by the level sets

$$D_\epsilon(X) = \cup_{s \in \Delta_\epsilon^*} X_s, \qquad (4.11)$$

where $X_s = \{t | f(t) = s\}$ and $\Delta_\epsilon^* \subset \mathbb{C}^*$ is a small punctured disk of radius $\epsilon > 0$. The boundary $\partial D_\epsilon(X)$ is given by

$$\partial D_\epsilon(X) = \cup_{s \in \partial \Delta_\epsilon^*} X_s. \qquad (4.12)$$

It is a circle bundle over the generic fiber X_ϵ, with projection

$$\pi_\epsilon : \partial D_\epsilon(X) \to X_\epsilon.$$

Given a domain of integration σ, possibly with boundary, we consider the intersection $\sigma \cap D_\epsilon(X)$. This contains the locus $\sigma \cap X$ which is of interest to us in terms of divergences in the Feynman integral. We let $\mathcal{L}_\epsilon(\sigma)$ denote the set

$$\mathcal{L}_\epsilon(\sigma) = \pi_\epsilon^{-1}(\sigma \cap X_\epsilon). \qquad (4.13)$$

This satisfies $\mathcal{L}_\epsilon(\partial\sigma) = \partial \mathcal{L}_\epsilon(\sigma)$.

We consider forms

$$\pi^*(\eta_m) = \frac{\Delta(\omega)}{f^m}, \qquad (4.14)$$

as in Proposition 3.4.3, where $X = \{f = 0\}$ with order of vanishing $m = -n + D(\ell+1)/2$, and $\omega = P_\Gamma(t, p)^{-n+D\ell/2}\omega_n$, with ω_n the standard volume form. We can then regularize the Feynman integral, written in the form of Proposition 3.4.3, in the following way.

Definition 4.2.3. The Leray regularized Feynman integral is obtained from (3.56) by replacing the part

$$\int_{\partial\sigma_n \cap D_\epsilon(X_\Gamma)} \pi^*(\eta_m) + \int_{\sigma_n \cap D_\epsilon(X_\Gamma)} df \wedge \frac{\pi^*(\eta_m)}{f} \qquad (4.15)$$

of (3.56) with the integral

$$\int_{\mathcal{L}_\epsilon(\partial\sigma_n)} \frac{\pi^*(\eta_{m-1})}{f - \epsilon} + \int_{\mathcal{L}_\epsilon(\sigma_n)} df \wedge \frac{\pi^*(\eta_m)}{f - \epsilon}, \qquad (4.16)$$

where $f = \Psi_\Gamma$ and $m = -n + D(\ell+1)/2$.

Thus, the Leray regularization introduced here consists in replacing the integral over $\sigma_n \cap D_\epsilon(X_\Gamma)$ with an integral over $\mathcal{L}_\epsilon(\sigma_n) \simeq (\sigma_n \cap X_{\Gamma,\epsilon}) \times S^1$, which avoids the locus $\sigma_n \cap X_\Gamma$ where the divergence can occur by going around it along a circle of small radius $\epsilon > 0$. Here $X_{\Gamma,\epsilon}$ is the ϵ-level set of $f = \Psi_\Gamma$.

Lemma 4.2.4. *The Leray regularization of the Feynman integral (3.56) can be equivalently written in the form*

$$U(\Gamma)_\epsilon = \frac{1}{C(n,D,\ell)} \left(\int_{\partial\sigma_n \cap D_\epsilon(X_\Gamma)^c} \pi^*(\eta_m) + \int_{\sigma_n \cap D_\epsilon(X_\Gamma)^c} df \wedge \frac{\pi^*(\eta_m)}{f} \right)$$

$$+ \frac{2\pi i}{C(n,D,\ell)} \left(\int_{\partial\sigma_n \cap X_{\Gamma,\epsilon}} \frac{\pi^*(\eta_{m-1})}{df} + \int_{\sigma_n \cap X_{\Gamma,\epsilon}} \pi^*(\eta_m) \right),$$

(4.17)

with $\pi^*(\eta_m) = \Delta(\omega)/f^m$ as in (4.14) and Proposition 3.4.3 and $D_\epsilon(X_\Gamma)^c$ denoting the complement of $D_\epsilon(X_\Gamma)$.

Proof. The result follows directly from Proposition 3.4.3 and Lemma 4.2.2 applied to (4.16). □

We now study the dependence on the parameter $\epsilon > 0$ of the Leray regularized Feynman integral (4.16), *i.e.* of an integral of the form

$$I_\epsilon := \int_{\partial\sigma \cap X_\epsilon} \frac{\pi^*(\eta_{m-1})}{df} + \int_{\sigma \cap X_\epsilon} \pi^*(\eta_m). \qquad (4.18)$$

Theorem 4.2.5. *The function I_ϵ of (4.18) is infinitely differentiable in ϵ. Moreover, it extends to a holomorphic function for $\epsilon \in \Delta^* \subset \mathbb{C}$, a small punctured disk, with a pole of order at most m at $\epsilon = 0$, with $m = -n + D(\ell+1)/2$.*

Proof. To prove the differentiability of I_ϵ, let us write

$$A_\epsilon(\eta) = \int_{\sigma \cap X_\epsilon} \pi^*(\eta), \qquad (4.19)$$

with $\pi^*(\eta)$ as in (3.57). By Lemma 4.2.2 above, and the fact that $d\pi^*(\eta) = 0$, we obtain

$$\frac{d}{d\epsilon} A_\epsilon(\eta) = - \int_{\partial\sigma \cap X_\epsilon} \frac{\pi^*(\eta)}{df}, \qquad (4.20)$$

where $X_\epsilon = \{f = \epsilon\}$ and $\pi^*(\eta)/df$ is the Gelfand–Leray form of $\pi^*(\eta)$. Thus, we can write

$$I_\epsilon = A_\epsilon(\eta_m) - \frac{d}{d\epsilon} A_\epsilon(\eta_{m-1}).$$

Thus, to check the differentiability in the variable ϵ to all orders of I_ϵ is equivalent to checking that of A_ϵ. We then define $\Upsilon : \Omega^n \to \Omega^n$ by setting

$$\Upsilon(\alpha) = d\left(\frac{\alpha}{df}\right), \tag{4.21}$$

where α/df is a Gelfand–Leray form for α. In turn, the n-form $\Upsilon(\alpha)$ also has a Gelfand–Leray form, which we denote by

$$\delta(\alpha) = \frac{\Upsilon(\alpha)}{df} = \frac{d\left(\frac{\alpha}{df}\right)}{df}. \tag{4.22}$$

We then prove that, for $k \geq 2$,

$$\frac{d^k}{d\epsilon^k} A_\epsilon = -\int_{\partial\sigma \cap X_\epsilon} \delta^{k-1}\left(\frac{\pi^*(\eta)}{df}\right). \tag{4.23}$$

This follows by induction. In fact, we first see that

$$\frac{d^2}{d\epsilon^2} A_\epsilon = -\frac{d}{d\epsilon}\int_{\partial\sigma \cap X_\epsilon} \frac{\pi^*(\eta)}{df}$$

which, applying Lemma 4.2.2 gives

$$= -\int_{\partial\sigma \cap X_\epsilon} \frac{d\left(\frac{\pi^*(\eta)}{df}\right)}{df}.$$

Assuming then that

$$\frac{d^k}{d\epsilon^k} A_\epsilon = -\int_{\partial\sigma \cap X_\epsilon} \delta^{k-1}\left(\frac{\pi^*(\eta)}{df}\right)$$

we obtain again by a direct application of Lemma 4.2.2

$$\frac{d^{k+1}}{d\epsilon^{k+1}} A_\epsilon = -\int_{\partial\sigma \cap X_\epsilon} \frac{d\left(\frac{\delta^{k-1}\left(\frac{\pi^*(\eta)}{df}\right)}{df}\right)}{df}$$

$$= -\int_{\partial\sigma \cap X_\epsilon} \delta^k\left(\frac{\pi^*(\eta)}{df}\right).$$

This proves differentiability to all orders.

Notice then that, while the expression (4.16) used in Definition 4.2.3 is, a priori, only defined for $\epsilon > 0$, the equivalent expression given in the second line of (4.17) and in (4.18) is clearly defined for any complex $\epsilon \in \Delta^*$ in a punctured disk around $\epsilon = 0$ of sufficiently small radius. It can then be seen that the expression (4.18) depends holomorphically on the parameter ϵ by the general argument on holomorphic dependence on parameters given in Part III, §10.2 of Vol.II of [Arnold, Gusein-Zade, Varchenko (1988)].

Finally, to see that I_ϵ has a pole of order at most m at $\epsilon = 0$, notice that the form $\pi^*(\eta_m)$ of (4.14) is given by $\Delta(\omega)/f^m$ and has a pole of order at most m at X.   □

Notice that the regularization procedure described here is natural from the point of view of singularity theory and oscillatory integrals, but cannot be used directly to derive information at the motivic level. To be able to do so one would have to first derive an algebraic version of the procedure described here, possibly involving a motivic version of the notion of "tubular neighborhood", see *e.g.* [Levine (2005)], applied to the locus $\Sigma \cap X_\Gamma$, with Σ a normal crossings divisor containing the boundary $\partial \sigma_n$ of the domain of integration, such as $\Sigma = \Sigma_n$ the union of the coordinate hyperplanes. Algebro-geometric notions of tubular neighborhood were also elaborated upon in Grothendieck's "Esquisse d'un Programme". The role of such notion in the algebro-geometric theory of Feynman integrals awaits further investigation.

Chapter 5

Connes–Kreimer theory in a nutshell

So far we focused on individual Feynman graphs and associated algebro-geometric constructions. However, it is well known in quantum field theory that the process of removing divergences from Feynman integrals via renormalization cannot be achieved without taking into account the nested structure of divergences, *i.e.* the presence of subgraphs inside a given Feynman graph that already contribute divergences to the Feynman integral. A process of subtracting divergences that takes into account this hierarchy of nested subdivergences was developed in [Bogolyubov and Parasiuk (1957)] and later completed by [Hepp (1966)] and [Zimmermann (1969)].

A very elegant geometric formulation of the BPHZ renormalization procedure was given in [Connes and Kreimer (2000)] and [Connes and Kreimer (2001)] in terms of a Hopf algebra of Feynman graphs, where the coproduct encodes the information on the subdivergences, and the BPHZ procedure becomes an extraction of finite values via the Birkhoff factorization of loops in a pro-unipotent Lie group of characters of the Hopf algebra of Feynman graphs. In this chapter we give a very brief account of the main steps in the Connes–Kreimer theory. For brevity, we skip several important technical points, such as how to deal with external momenta and we also do not reproduce full proofs of the main results. In fact, these are already described in full detail in the original papers of Connes and Kreimer and also in the first chapter of the book [Connes and Marcolli (2008)], so we do not duplicate them here. We restrict to the essential skeleton of what we need in order to continue with our narrative in the following chapters.

5.1   The Bogolyubov recursion

The main steps of the Bogolyubov–Parasiuk–Hepp–Zimmermann renormalization procedure (BPHZ) are summarized as follows. (Again, for more
details the reader is invited to look at Chapter 1 on [Connes and Marcolli
(2008)]).

5.1.1   *Step 1: Preparation*

One replaces the Feynman integral $U(\Gamma)$ of (1.55) by the expression

$$\bar{R}(\Gamma) = U(\Gamma) + \sum_{\gamma \in \mathcal{V}(\Gamma)} C(\gamma) U(\Gamma/\gamma). \qquad (5.1)$$

Here we suppress the dependence on z, μ and the external momenta p for
simplicity of notation. The expression (5.1) is to be understood as a sum
of Laurent series in z, depending on the extra parameter μ. The coeffcients
$C(\gamma)$, themselves Laurent series, are defined recursively in (5.2) below.

The sum in (5.1) is over the set $\mathcal{V}(\Gamma)$ of (not necessarily connected)
proper non-empty subgraphs $\gamma \subset \Gamma$ with the property that the *residue*
$\mathrm{Res}(\gamma_i)$ of each component γ_i of the graph γ has valence equal to one of the
monomials in the Lagrangian $\mathcal{L}(\phi)$. Here the residue is defined as follows.

Definition 5.1.1. The *residue* $\mathrm{Res}(\Gamma)$ of a connected Feynman graph Γ is
the Feynman graph consisting of a single vertex and as many half edges
attached to it as the number of external edges of Γ, *i.e.* the graph obtained
by contracting all the internal edges of Γ to a single vertex.

5.1.2   *Step 2: Counterterms*

These are the expressions by which the Lagrangian needs to be modified
to cancel the divergence produced by the graph Γ. They are defined as the
polar part of the Laurent series $\bar{R}(\Gamma)$,

$$C(\Gamma) = -\mathfrak{T}(\bar{R}(\Gamma)). \qquad (5.2)$$

Here $\mathfrak{T}$ denotes the operator of projection of Laurent series onto their polar
part. Notice how the counterterms $C(\Gamma)$ are defined here by a recursion,
where in the right hand side of (5.2) only terms $C(\gamma)$ for strictly smaller
graphs $\gamma \subset \Gamma$ are involved.

We discuss here briefly a property of the operator $\mathfrak{T}$ that will be useful
later. We'll see in §5.5 below that this property corresponds to the fact

that the differential field $\mathcal{K}$ of convergent Laurent series with the operator $\mathfrak{T}$ has the structure of a Rota–Baxter algebra.

Lemma 5.1.2. *Let $\mathcal{K}$ be the field of convergent Laurent series (germs of meromorphic functions at the origin). Let $\mathfrak{T}$ be, as above, the operator that projects a Laurent series onto its polar part. This satisfies the identity*

$$\mathfrak{T}(f_1)\mathfrak{T}(f_2) = \mathfrak{T}(f_1\mathfrak{T}(f_2)) + \mathfrak{T}(\mathfrak{T}(f_1)f_2) - \mathfrak{T}(f_1f_2). \tag{5.3}$$

Proof. If $f_1 = \sum_{k=-N_1}^{\infty} a_k z^k$ and $f_2 = \sum_{r=-N_2}^{\infty} b_r z^r$ then

$$\mathfrak{T}(f_1 f_2) = \sum_{m=-(N_1+N_2)}^{-1} \sum_{k+r=m} a_k b_r\, z^m.$$

On the other hand, we also have

$$\mathfrak{T}(f_1)\mathfrak{T}(f_2) = \Big(\sum_{k=-N_1}^{-1} a_k z^k \Big)\Big(\sum_{r=-N_2}^{-1} b_r z^r \Big).$$

One then finds the identity (5.3), since all the terms that appear in $\mathfrak{T}(f_1 f_2)$ are accounted for with multiplicity one by summing

$$\mathfrak{T}(\mathfrak{T}(f_1)f_2) = \mathfrak{T}\Big(\big(\sum_{k=-N_1}^{-1} a_k z^k \big)\big(\sum_{r=-N_2}^{\infty} b_r z^r \big) \Big)$$

and

$$\mathfrak{T}(f_1\mathfrak{T}(f_2)) = \mathfrak{T}\Big(\big(\sum_{k=-N_1}^{\infty} a_k z^k \big)\big(\sum_{r=-N_2}^{-1} b_r z^r \big) \Big)$$

and then subtracting $\mathfrak{T}(f_1)\mathfrak{T}(f_2)$, which has the effect of removing the terms that had been counted twice in the previous sum and leaving exactly the terms that constitute the polar part $\mathfrak{T}(f_1 f_2)$. $\qquad\square$

5.1.3 Step 3: Renormalized values

One can then extract a finite value from the integral $U(\Gamma)$ by removing the polar part, not of $U(\Gamma)$ itself but of its preparation. Namely, one sets

$$R(\Gamma) = \bar{R}(\Gamma) + C(\Gamma) = U(\Gamma) + C(\Gamma) + \sum_{\gamma \in \mathcal{V}(\Gamma)} C(\gamma)U(\Gamma/\gamma), \tag{5.4}$$

which we also write equivalently as

$$R(\Gamma) = U(\Gamma) + \sum_{\gamma \in \mathcal{V}(\Gamma)} C(\gamma)U(\Gamma/\gamma) - \mathfrak{T}\left(U(\Gamma) + \sum_{\gamma \in \mathcal{V}(\Gamma)} C(\gamma)U(\Gamma/\gamma) \right),$$

$$\tag{5.5}$$

that is, the minimal subtraction of the polar part not of the original Laurent series $U(\Gamma)$ but of the corrected one $\bar{R}(\Gamma)$.

The main result of BPHZ is that, unlike what happens typically with the Laurent series $U(\Gamma)$ alone, the coefficient of pole of the corrected Laurent series $\bar{R}(\Gamma)$ provided by the Bogolyubov preparation is always a local expression in the variables of the external momenta, hence one that can be corrected either by altering the coefficients of the already existing terms in the Lagrangian or by a adding a *finite number* of new polynomial terms to the Lagrangian. Namely, the following statement is the final result of the combined [Bogolyubov and Parasiuk (1957)], [Hepp (1966)] and [Zimmermann (1969)].

Theorem 5.1.3. *The coefficient of pole of $\bar{R}(\Gamma)$ is local.*

Here local means that it can be expressed as a local function in the momentum variables (involving only polynomials and derivatives). In fact, if one wants to be able to remove the polar part by correcting the Lagrangian by adding suitable counterterms, these themselves have to be local functions as the terms already present in the Lagrangian are.

The BPHZ procedure also goes under the name of *Bogolyubov recursion*, because of the recursive nature of the definition of the counterterms $C(\Gamma)$ in (5.2). A very nice conceptual understanding of the BPHZ renormalization procedure with the DimReg+MS regularization was obtained by Connes and Kreimer [Connes and Kreimer (2000)], [Connes and Kreimer (2001)], based on a reformulation of the BPHZ procedure in geometric terms. A detailed account of the Connes–Kreimer theory was given in Chapter 1 of [Connes and Marcolli (2008)], so we limit ourselves to a more sketchy overview here and refer the reader to that more detailed treatment.

5.2 Hopf algebras and affine group schemes

In the following we work with algebras defined over a field $\mathbb{K}$ of characteristic zero. In fact, for our main application, $\mathbb{K}$ will be either $\mathbb{Q}$ or $\mathbb{C}$. We also restrict our attention here to commutative Hopf algebras, because the Hopf algebras that appear in applications to the theory of renormalization in quantum field theory are usually of this kind.

Definition 5.2.1. A *commutative Hopf algebra* over $\mathbb{K}$ is a commutative $\mathbb{K}$-algebra $\mathcal{H}$ endowed with an algebra homomorphism $\Delta : \mathcal{H} \to \mathcal{H} \otimes \mathcal{H}$, the *coproduct*, an algebra homomorphism $\varepsilon : \mathcal{H} \to \mathbb{K}$, the counit, and an algebra antihomomorphism $S : \mathcal{H} \to \mathcal{H}$, the antipode. These satisfy the coassociativity

$$(\Delta \otimes id)\Delta = (id \otimes \Delta)\Delta$$

as morphisms $\mathcal{H} \to \mathcal{H} \otimes_{\mathbb{K}} \mathcal{H} \otimes_{\mathbb{K}} \mathcal{H}$, the compatibility of the coproduct Δ with the counit ε,

$$(id \otimes \varepsilon)\Delta = id = (\varepsilon \otimes id)\Delta$$

as morphisms $\mathcal{H} \to \mathcal{H}$, and the compatibility of comultiplication Δ, antipode S, and multiplication μ,

$$\mu(id \otimes S)\Delta = \mu(S \otimes id)\Delta = 1\,\varepsilon$$

as morphisms $\mathcal{H} \to \mathcal{H}$.

Notice that in general the coproduct will not be cocommutative.

A useful notion we refer to often in the following, which also already appeared in §2.9 and §2.10, in the context of the Tannakian formalism and motivic Galois groups, is that of affine group scheme. As we recall briefly here (see also [Waterhouse (1979)]) this is a functor from algebras to groups naturally associated to a commutative Hopf algebra.

Proposition 5.2.2. *A commutative Hopf algebra $\mathcal{H}$ over a field $\mathbb{K}$ of characteristic zero determines a covariant functor G from the category $\mathcal{A}_{\mathbb{K}}$ of commutative algebras over $\mathbb{K}$ to the category of groups, by setting*

$$G(A) = \mathrm{Hom}_{\mathcal{A}_{\mathbb{K}}}(\mathcal{H}, A), \tag{5.6}$$

for $A \in \mathrm{Obj}(\mathcal{A}_{\mathbb{K}})$.

Proof. We need to show that the set $G(A)$ is a group. Elements of $G(A)$ are homomorphisms of $\mathbb{K}$-algebras

$$\phi : \mathcal{H} \to A, \quad \phi(x\,y) = \phi(x)\phi(y), \quad \forall x, y \in \mathcal{H}, \quad \phi(1) = 1\,.$$

There is a product operation on $G(A)$ defined by the coproduct Δ of $\mathcal{H}$

$$\phi_1 * \phi_2\,(x) = \langle \phi_1 \otimes \phi_2\,, \Delta(x)\rangle\,. \tag{5.7}$$

It is associative because of the coassociativity of Δ. There is a unit element in $G(A)$ which is determined by the counit ε of $\mathcal{H}$ by

$$1_{G(A)} : \mathcal{H} \xrightarrow{\varepsilon} \mathbb{K} \to A.$$

The property that this is a unit with respect to the multiplication in $G(A)$ come from the compatibility relation between counit and coproduct in the axioms of Hopf algebra. Similarly, there is an inverse $\phi^{-1} = \phi \circ S$, defined by pre-composing a morphism $\phi : \mathcal{H} \to A$ with the antipode $S : \mathcal{H} \to \mathcal{H}$. The fact that it satisfies the correct properties with respect to the multiplication and unit,

$$\phi * \phi^{-1} = \phi^{-1} * \phi = 1,$$

to be the inverse in the group $G(A)$ follows from the remaining axiom of the Hopf algebra structure that gives the compatibility between the counit, co-multiplication, and antipode. The map $\phi \mapsto \phi^{-1}$ is an anti-homomorphism, $(\phi_1 * \phi_2)^{-1} = \phi_2^{-1} * \phi_1^{-1}$, because S is an anti-homomorphism of the Hopf algebra $\mathcal{H}$. $\square$

An *affine group scheme* is a representable covariant functor from the category of commutative algebras over the field $\mathbb{K}$ to the category of groups. So the above shows that a commutative Hopf algebra determines an affine group scheme. Any such representable functor will be of the form (5.6), for some commutative algebra $\mathcal{H}$, which in turn inherits the structure of Hopf algebra from the group structure of $G(A)$.

The simplest examples of affine group schemes are the additive and the multiplicative group, $\mathbb{G}_a$ and $\mathbb{G}_m$, respectively, which correspond to the Hopf algebras

- The additive group $\mathbb{G}_a$ corresponds to the Hopf algebra $\mathcal{H} = \mathbb{K}[t]$ with coproduct $\Delta(t) = t \otimes 1 + 1 \otimes t$, counit $\varepsilon(t) = 0$ and antipode $S(t) = -t$.
- The multiplicative group $\mathbb{G}_m$ corresponds to the Hopf algebra $\mathcal{H} = \mathbb{K}[t, t^{-1}]$, with coproduct $\Delta(t) = t \otimes t$, counit $\varepsilon(t) = 1$ and antipode $S(t) = t^{-1}$.

The main examples in the following will be pro-unipotent affine group schemes. These are projective limits of unipotent algebraic groups, where by algebraic groups we mean here subgroups of the affine group scheme GL_n. These pro-unipotent cases arise from Hopf algebras that are graded, $\mathcal{H} = \oplus_{n \geq 0} \mathcal{H}_n$ and connected, $\mathcal{H}_0 = \mathbb{K}$, as the Connes–Kreimer Hopf algebra is. We discuss more the example of GL_n as an affine group scheme in Proposition 5.4.2 below. For more detailed information on affine group schemes, we refer the reader to [Demazure and Grothendieck (1970)] and [Waterhouse (1979)].

5.3   The Connes–Kreimer Hopf algebra

The Hopf algebra of Feynman graphs introduced in [Connes and Kreimer (2000)] depends on the choice of the physical theory, in the sense that it involves only those graphs that are Feynman graphs for the specified Lagrangian $\mathcal{L}(\phi)$, and, as we see below, the coproduct formula also involves only those subgraphs whose residue graph corresponds to one of the monomials in the Lagrangian.

As an algebra $\mathcal{H}_{CK}$ it is the free commutative algebra with generators the 1PI Feynman graphs Γ of the specified physical theory. It is graded by loop number, or by the number of internal lines, with

$$\deg(\Gamma_1 \cdots \Gamma_n) = \sum_i \deg(\Gamma_i), \quad \deg(1) = 0.$$

This grading corresponds to the order in the perturbative expansion.

The most interesting part of the structure is the coproduct, which is given on generators by

$$\Delta(\Gamma) = \Gamma \otimes 1 + 1 \otimes \Gamma + \sum_{\gamma \in \mathcal{V}(\Gamma)} \gamma \otimes \Gamma/\gamma. \tag{5.8}$$

The sum is over the class $\mathcal{V}(\Gamma)$ of proper subgraphs of Γ with the property that the quotient graph Γ/γ is still a 1PI Feynman graph of the same theory.

The subgraphs in $\mathcal{V}(\Gamma)$ are not necessarily connected. In the case of several connected components, the quotient graph Γ/γ is understood as the graph obtained by shrinking each component of $\gamma \subset \Gamma$ to a vertex.

The condition that Γ/γ is a 1PI Feynman graph can be formulated in terms of residue graphs $\mathrm{Res}(\gamma)$ of the sugraphs $\gamma \subset \Gamma$. For a given graph Γ the residue $\mathrm{Res}(\Gamma)$ is the graph that is obtained by keeping all the external edges of Γ and shrinking all internal edges to a single vertex. The valence of this vertex is then equal to the number of external edges. The class of subgraphs $\mathcal{V}(\Gamma)$ can then be charcterized by requiring that the components of $\gamma \subset \Gamma$ are 1PI Feynman graphs with the property that the residue $\mathrm{Res}(\Gamma)$ has valence corresponding to one of the monomials in the Lagrangian of the theory. This in fact ensures that the corresponding vertex in the quotient graph Γ/γ has an admissible valence for it to be a Feynman graph. The fact that Γ/γ is 1PI then follows from the fact that Γ and all the components of γ are 1PI.

The antipode is defined inductively by

$$S(X) = -X - \sum S(X')X'',$$

where X is an element with coproduct $\Delta(X) = X \otimes 1 + 1 \otimes X + \sum X' \otimes X''$, where all the X' and X'' have lower degrees.

The affine group scheme dual to the Connes–Kreimer Hopf algebra is referred to as the *group of diffeographisms*. Its set of complex points

$$G(\mathbb{C}) = \mathrm{Hom}(\mathcal{H}_{CK}, \mathbb{C})$$

is a pro-unipotent complex Lie group $G(\mathbb{C})$. It was proved in [Connes and Kreimer (2000)] that this group acts by local diffeomorphisms on the coupling constants of the theory.

A variant of the Connes–Kreimer Hopf algebra, which is useful in algebro-geometric applications, was considered in [Bloch and Kreimer (2008)], see also [Kreimer (2009)]. It is referred to as the *core Hopf algebra* and it is simply defined by considering as generators all 1PI graphs without restriction on the valence of vertices. This would correspond in physical terms to arbitrarily large powers ϕ^k in the Lagrangian. Correspondingly, the coproduct is taken over all the subgraphs whose connected components are 1PI, without restriction on the residue graphs.

The algebro-geometric Feynman rules discussed above and introduced in [Aluffi and Marcolli (2008b)] define ring homomorphisms from the core Hopf algebra to the Grothendieck ring of varieties, or to the ring of immersed conical varieties defined in [Aluffi and Marcolli (2008b)].

A further important remark about the Connes–Kreimer Hopf algebra: the version we recalled briefly here is only the combinatorial part of a larger Hopf algebra, whose generators are pairs (Γ, σ) of a 1PI graph and a distribution on the space of test functions of the external momenta. This continuous version of the the Hopf algebra of renormalization is necessary when taking into account the external momenta of the graphs in the extraction of subdivergences. This additional structure is described in detail in [Connes and Kreimer (2000)] as well as in Chapter 1 of [Connes and Marcolli (2008)].

Finally, the Connes–Kreimer construction of the Hopf algebra is given for scalar quantum field theories. Generalizations to vector valued fields do not present any problem, while extending the same setting to gauge theories and chiral theories is more subtle. The results were recently generalized to gauge theories in [van Suijlekom (2007)], [van Suijlekom (2006)], [van Suijlekom (2008)], by showing that the Ward identities define Hopf ideals. The case of chiral theories runs into a well known problem of reconciling the use of dimensional regularization with the chirality operator γ_5, see [Jegerlehner (2001)]. A way to describe the prescription of [Breitenlohner and Maison

(1977)] to express the chirality operator γ_5 within DimReg in terms of noncommutative geometry was given in the unpublished [Connes and Marcolli (2005b)]; see also Chapter 1 of [Connes and Marcolli (2008)]. In this setting the operation of dimensional regularization is then expressed as a cup product of spectral triples. Recently, a version of the Connes–Kreimer renormalization as Birkhoff factorization, based on zeta function regularization instead of DimReg and adapted to curved backgrounds was developed in [Agarwala (2009)]. An extension of the parametric Feynman integrals to theories with bosonic and fermionic fields, using periods of supermanifolds, was given in [Marcolli and Rej (2008)], though the most appropriate form of the Connes–Kreimer renormalization procedure in that context remains to be properly discussed. Another interesting recent result is a categorification of the Connes–Kreimer Hopf algebra obtained in [Kremnizer and Szczesny (2008)], where the dual Hopf algebra is identified with the Hall algebra of a finitary abelian category.

5.4 Birkhoff factorization

We give an algebraic formulation of Birkhoff factorization in the group scheme G of a commutative Hopf algebra in terms of the Hopf algebra $\mathcal{H}$ and its characters.

Definition 5.4.1. Suppose given an *algebra homomorphism* $\phi \in$ $\mathrm{Hom}(\mathcal{H}, \mathcal{K})$, where $\mathcal{K}$ is the field of convergent Laurent series, *i.e.* germs of meromorphic functions at $z = 0$. Then ϕ has a Birkhoff factorization if there exist *algebra homomorphisms* $\phi_+ \in \mathrm{Hom}(\mathcal{H}, \mathcal{O})$, with $\mathcal{O}$ the complex algebra of convergent power series, and $\phi_- \in \mathrm{Hom}(\mathcal{H}, \mathcal{Q})$, with values in the polynomial algebra $\mathcal{Q} = \mathbb{C}[z^{-1}]$, such that

$$\phi = (\phi_- \circ S) \star \phi_+, \qquad (5.9)$$

where S is the antipode in the Hopf algebra and $\star$ is the product in G dual to the coproduct in the Hopf algebra. The factorization (5.9), if it exists, is not unique. To make it unique one also requires the normalization condition

$$\epsilon_- \circ \phi_- = \epsilon, \qquad (5.10)$$

where $\epsilon_- : \mathbb{C}[z^{-1}] \to \mathbb{C}$ is the augmentation and ϵ is the counit of the Hopf algebra $\mathcal{H}$.

The characters $\phi \in \mathrm{Hom}(\mathcal{H}, \mathcal{K})$ can be equally interpreted as *loops* defined on an infinitesimal punctured disk Δ^* around the origin $z = 0$ in $\mathbb{C}^*$ and with values in the prounipotent complex Lie group $G(\mathbb{C}) = \mathrm{Hom}(\mathcal{H}, \mathbb{C})$ of algebra homomorphisms from $\mathcal{H}$ to $\mathbb{C}$, so that the recursive formula for the factorization of ϕ gives the Birkhoff factorization of loops written in the more traditional form

$$\gamma(z) = \gamma_-(z)^{-1}\gamma_+(z), \qquad (5.11)$$

with the normalization condition $\gamma_-(\infty) = 1$.

Not all commutative Hopf algebras admit a Birkhoff factorization for arbitrary elements $\phi \in \mathrm{Hom}(\mathcal{H}, \mathcal{K})$. The following example illustrates a case where there are elements in $\mathrm{Hom}(\mathcal{H}, \mathcal{K})$ that do not admit a factorization of the form (5.9).

Proposition 5.4.2. *Consider the commutative Hopf algebra* $\mathcal{H} = \mathbb{C}[x_{ij}, t]/(\det(x_{ij}) - t)$ *with the coproduct*

$$\Delta(x_{ij}) = \sum_k x_{ik} \otimes x_{kj}.$$

Then there exist elements $\phi \in \mathrm{Hom}(\mathcal{H}, \mathcal{K})$ *that do not admit a factorization of the form* (5.9).

Proof. The affine group scheme dual to the Hopf algebra $\mathcal{H} = \mathbb{C}[x_{ij}, t]/(\det(x_{ij}) - t)$ with the given coproduct is the group GL_n. To see that there exist elements of $\phi \in \mathrm{Hom}(\mathcal{H}, \mathcal{K})$ that do not admit a Birkhoff factorization in the form stated above, we can give an equivalent formulation in terms of loops in a complex Lie group. We can use the equivalent description of characters $\phi \in \mathrm{Hom}(\mathcal{H}, \mathcal{K})$ in terms of loops $\gamma : \Delta^* \to \mathrm{GL}_n$, as above. Then, it suffices to notice that, if we use the function $\gamma(z)$ to define the transition function of a holomorphic vector bundle over the 2-sphere $\mathbb{P}^1(\mathbb{C})$, the existence of a factorization (5.11) shows that the resulting bundle is trivial. However, one knows that there are non-trivial holomorphic vector bundles on $\mathbb{P}^1(\mathbb{C})$, so it suffices to take one such non-trivial vector bundle E and a transition function $\gamma(z)$ relating the trivializations of E on the two contractible sets Δ and $\mathbb{P}^1(\mathbb{C}) \smallsetminus \Delta$ to see that this loop cannot admit a factorization of the form (5.11). $\qquad \square$

One of the main results of the Connes–Kreimer theory is a recursive formula for the Birkhoff factorization in the case of graded connected commutative Hopf algebras. We review this result in the next section.

5.5 Factorization and Rota-Baxter algebras

Let $\mathcal{H}$ be a graded connected commutative Hopf algebra $\mathcal{H} = \oplus_{n \geq 0} \mathcal{H}_n$, where the $\mathcal{H}_n$ are finite dimensional. The connectedness condition means that $\mathcal{H}_0 = \mathbb{C}$, for an algebra defined over the field of complex numbers. Notice that here the condition that $\mathcal{H}$ is a commutative graded Hopf algebra does not mean that it is "graded-commutative": the commutators are not graded commutators but ordinary commutators. It behaves like the case of a graded–commutative Hopf algebra where the graded pieces are non-zero only in even degrees.

Let $\mathcal{K}$ denote the field of convergent Laurent series. We consider, as above, elements $\phi \in \mathrm{Hom}(\mathcal{H}, \mathcal{K})$ and we ask whether all such elements admit a Birkhoff factorization as in (5.9). The Connes–Kreimer formula [Connes and Kreimer (2000)] gives a positive answer, for the class of Hopf algebras mentioned above, in the form of an explicit recursive formula. In fact, for later use, we can formulate their result in a slightly more general form, as was done for instance in [Ebrahimi-Fard, Guo, Kreimer (2004)], using the formalism of Rota–Baxter algebras.

To this purpose we recall some preliminary notions on Rota–Baxter algebras and operators.

Definition 5.5.1. A commutative Rota–Baxter algebra of weight λ is a commutative algebra $\mathcal{A}$ endowed with a linear operator $\mathfrak{R}$ satisfying the Rota–Baxter identity

$$\mathfrak{R}(x)\mathfrak{R}(y) = \mathfrak{R}(x\mathfrak{R}(y)) + \mathfrak{R}(\mathfrak{R}(x)y) + \lambda\mathfrak{R}(xy). \tag{5.12}$$

Such an operator is called a Rota–Baxter operator of weight λ.

The prototype of a Rota–Baxter operator is integration, for which the Rota–Baxter identity is the integration by parts formula.

Lemma 5.5.2. *The pair* $(\mathcal{K}, \mathfrak{T})$ *of the field of convergent Laurent series with the projection on the polar part is a Rota–Baxter algebra of weight* $\lambda = -1$.

Proof. This is an immediate consequence of Lemma 5.1.2. $\qquad\square$

Lemma 5.5.3. *Given a commutative unital Rota–Baxter algebra* $(\mathcal{A}, \mathfrak{R})$ *of weight* $\lambda = -1$, *one obtains two commutative unital algebras* $\mathcal{A}_{\pm}$ *by taking as* $\mathcal{A}_-$ *the image* $\mathfrak{R}(\mathcal{A})$ *with a unit adjoined (i.e. the smallest unital algebra containing* $\mathfrak{R}(\mathcal{A})$ *as an ideal) and* $\mathcal{A}_+ = (1 - \mathfrak{R})(\mathcal{A})$.

Proof.   These are indeed algebras. In fact, for $x, y \in \mathcal{A}_-$ we have $x = \mathfrak{R}(a)$ and $y = \mathfrak{R}(b)$ for some $a, b \in \mathcal{A}$. The product $\mathfrak{R}(a)\mathfrak{R}(b) = \mathfrak{R}(a\mathfrak{R}(b) + \mathfrak{R}(a)b + \lambda ab)$ is also in the range of $\mathfrak{R}$, hence in $\mathcal{A}_-$. Similarly, given $x, y \in \mathcal{A}_+$ we have $x = a - \mathfrak{R}(a)$ and $y = b - \mathfrak{R}(b)$ for some $a, b \in \mathcal{A}$. Then

$$xy = (a - \mathfrak{R}(a))(b - \mathfrak{R}(b)) = ab - a\mathfrak{R}(b) - b\mathfrak{R}(a) + \mathfrak{R}(a)\mathfrak{R}(b)$$

$$= ab - a\mathfrak{R}(b) - b\mathfrak{R}(a) + \mathfrak{R}(a\mathfrak{R}(b) + \mathfrak{R}(a)b + \lambda ab).$$

Thus, for $\lambda = -1$, one obtains

$$xy = (ab - a\mathfrak{R}(b) - b\mathfrak{R}(a)) - \mathfrak{R}(ab - a\mathfrak{R}(b) - b\mathfrak{R}(a)) \in \mathcal{A}_+. \qquad \square$$

Just as Definition 5.4.1 formulates the Birkhoff factorization for algebra homomorphisms $\phi \in \mathrm{Hom}(\mathcal{H}, \mathcal{K})$ we can similarly consider the same problems for algebra homomorphisms from the Hopf algebra $\mathcal{H}$ with values in any Rota–Baxter algebra $(\mathcal{A}, \mathfrak{R})$ of weight $\lambda = -1$. Given $\phi \in \mathrm{Hom}(\mathcal{H}, \mathcal{A})$, we look for algebra homomorphisms $\phi_\pm \in \mathrm{Hom}(\mathcal{H}, \mathcal{A}_\pm)$ with the algebras $\mathcal{A}_\pm$ obtained as in Lemma 5.5.3, satisfying the factorization

$$\phi = (\phi_- \circ S) \star \phi_+, \tag{5.13}$$

as in (5.9), where here we use the fact that $\mathcal{A}_\pm \subset \mathcal{A}$ so that the equality above is understood as the equality

$$\phi(X) = (\phi_- \circ S)(X) \star \phi_+(X), \quad \forall X \in \mathcal{H}$$

viewed as an equality of elements in $\mathcal{A}$. The normalization condition is also formulated in this case as $\epsilon_- \circ \phi_- = \epsilon$, where $\epsilon_- : \mathcal{A}_- \to \mathbb{C}$ is the augmentation map and ϵ is the counit of the Hopf algebra $\mathcal{H}$.

It was proved in [Connes and Kreimer (2000)], and reformulated in the Rota-Baxter context in [Ebrahimi-Fard, Guo, Kreimer (2004)], that if $\mathcal{H}$ is a graded connected commutative Hopf algebra and $(\mathcal{A}, \mathfrak{R})$ is a unital Rota–Baxter algebra of weight $\lambda = -1$, then any homomorphism of unital commutative algebras $\phi \in \mathrm{Hom}(\mathcal{H}, \mathcal{A})$ admits a Birkhoff factorization (5.13) where the homomorphisms $\phi_\pm \in \mathrm{Hom}(\mathcal{H}, \mathcal{A}_\pm)$ are defined inductively by the formulae

$$\phi_-(X) = -\mathfrak{R}\left(\phi(X) + \sum \phi_-(X')\phi(X'')\right), \tag{5.14}$$

where $\Delta(X) = X \otimes 1 + 1 \otimes X + \sum X' \otimes X''$, with X' and X'' of lower degree, and

$$\phi_+(X) = (1 - \mathfrak{R})\left(\phi(X) + \sum \phi_-(X')\phi(X'')\right). \tag{5.15}$$

The Rota-Baxter property of the operator $\mathfrak{R}$ with weight -1 is needed to show that the $\phi_\pm$ defined recursively in this way are algebra homomorphisms, $\phi_\pm(XY) = \phi_\pm(X)\phi_\pm(Y)$.

Thus, the results of [Connes and Kreimer (2000)] and [Connes and Kreimer (2001)] show that, in terms of Birkhoff factorization, the formulae of the BPHZ renormalization procedure (5.2), (5.4), (5.5) are exactly the recursive formulae for the Birkhoff factorization of characters of the Connes–Kreimer Hopf algebra with values in a Rota–Baxter algebra, namely

$$U_-(\Gamma) = -\mathfrak{T}\left(U(\Gamma) + \sum_{\gamma \subset \Gamma} U_-(\gamma)U(\Gamma/\gamma)\right), \qquad (5.16)$$

$$U_+(\Gamma) = (1 - \mathfrak{T})\left(U(\Gamma) + \sum_{\gamma \subset \Gamma} U_-(\gamma)U(\Gamma/\gamma)\right). \qquad (5.17)$$

This is just the same as the formulae (5.14) and (5.14), given above for any algebra homomorphism from a graded connected commutative Hopf algebra $\mathcal{H}$ to a commutative Rota–Baxter algebra $(\mathcal{A}, \mathfrak{R})$ of weight -1, applied to the case where $\mathcal{H} = \mathcal{H}_{CK}$ is the Connes–Kreimer Hopf algebra and where the Rota-Baxter algebra is given by Laurent series in the DimReg variable z and $\mathfrak{R} = \mathfrak{T}$ is the projection onto the polar part, and the homomorphism U is the usual Feynman amplitude regularized using DimReg. Then the $C(\Gamma) = U_-(\Gamma)$ are the counterterms and the $R(\Gamma) = U_+(\Gamma)|_{z=0}$ are the renormalized values.

5.6 Motivic Feynman rules and Rota-Baxter structure

We have seen, following [Aluffi and Marcolli (2008b)], that one can define motivic Feynman rules in terms of the class $[\mathbb{A}^n \setminus \hat{X}_\Gamma]$ of the graph hypersurface complement in the Grothendieck ring of varieties $K_0(\mathcal{V})$.

A variant of such motivic Feynman rules is obtained by setting

$$U(\Gamma) = \frac{[\mathbb{A}^n \setminus \hat{X}_\Gamma]}{\mathbb{L}^n}, \qquad (5.18)$$

with values in the ring $K_0(\mathcal{V})[\mathbb{L}^{-1}]$, where one inverts the Lefschetz motive, or equivalently, via the additive invariant of [Gillet and Soulé (1996)], with values in the Grothendieck ring of motives $K_0(\mathcal{M})$. Dividing by $\mathbb{L}^n$ has the effect of normalizing the "motivic Feynman rule" $[\mathbb{A}^n \setminus \hat{X}_\Gamma]$ by the value it would have if Γ were a forest on the same number of edges. For the original Feynman integrals this would measure the amount of linear dependence between the edge momentum variables created by the presence of the interaction vertices.

In the ring $K_0(\mathcal{V})[\mathbb{L}^{-1}]$ or $K_0(\mathcal{M})$ one can still consider the Rota–Baxter structure given by the operator $\mathfrak{R}$ of projection onto the polar part in the variable $\mathbb{L}$. The renormalized Feynman rule

$$U_+(\Gamma) = (1 - \mathfrak{R}) \left(U(\Gamma) + \sum_{\gamma \subset \Gamma} U_-(\gamma) U(\Gamma/\gamma) \right)$$

then gives a class in the Grothendieck ring of varieties $K_0(\mathcal{V})$. More interestingly, the renormalized value of the Feynman integral, in this case, gives some information on the birational geometry of the graph hypersurface.

In fact, we first recall the following useful result of [Larsen and Lunts (2003)]. We work here with the Grothendieck ring $K_0(\mathcal{V}_{\mathbb{C}})$ for varieties over the field of complex numbers.

It is proved in [Larsen and Lunts (2003)] that the quotient of the Grothendieck ring $K_0(\mathcal{V}_{\mathbb{C}})$ by the ideal generated by $\mathbb{L} = [\mathbb{A}^1]$ is isomorphic as a ring to $\mathbb{Z}[SB]$, the ring of the multiplicative monoid SB of stable birational equivalence classes of varieties in $\mathcal{V}_{\mathbb{C}}$. Recall that two (irreducible) varieties X and Y are stably birationally equivalent if $X \times \mathbb{P}^n$ and $Y \times \mathbb{P}^m$ are birationally equivalent for some $n, m \geq 0$.

We then see that the *renormalized value* is in this case

$$U_+(\Gamma)|_{\mathbb{L}=0} = (1 - \mathfrak{R}) \left(U(\Gamma) + \sum_{\gamma \subset \Gamma} U_-(\gamma) U(\Gamma/\gamma) \right) |_{\mathbb{L}=0} \in \mathbb{Z}[SB]. \quad (5.19)$$

Notice that the parts of $[\mathbb{A}^n \setminus \hat{X}_\Gamma]$, $[\mathbb{A}^{n_1} \setminus \hat{X}_\gamma]$ and $[\mathbb{A}^{n_2} \setminus \hat{X}_{\Gamma/\gamma}]$ that are contained in the ideal $(\mathbb{L}) \subset K_0(\mathcal{V}_0)$ contribute cancellations to the $\mathbb{L}^n$ in the denominator. It is possible that this invariant and the Birkhoff factorization of $U(\Gamma)$ may help to detect the presence of non-mixed-Tate strata in the graph hypersurface X_Γ coming from the contributions of hypersurfaces of smaller graphs $\gamma \subset \Gamma$ or quotient graphs Γ/γ.

The class of a graph hypersurface X_Γ in the ring $\mathbb{Z}[SB]$ of stable birational equivalence classes detects some of the properties of the graph Γ. For example, it was shown in [Aluffi and Marcolli (2008b)] that the class is trivial, $[X_\Gamma]_{SB} = 1$, whenever the graph is not 1PI. In fact, if a graph is not 1PI then in the Grothendieck ring one has $[\mathbb{A}^n \setminus \hat{X}_\Gamma] = \mathbb{L} \cdot [\mathbb{A}^{n-1} \setminus \hat{X}_{\Gamma'}]$, where Γ' is the graph obtained by removing a disconnecting edge of Γ. Thus, $[\mathbb{A}^n \setminus \hat{X}_\Gamma]_{SB} = 0 \in \mathbb{Z}[SB]$. Using then the fact that

$[\mathbb{A}^n \smallsetminus \hat{X}_\Gamma] = \mathbb{L}^n - 1 - (\mathbb{L} - 1)[X_\Gamma]$ in the Grothendieck ring, one obtains $[\mathbb{A}^n \smallsetminus \hat{X}_\Gamma]_{\mathrm{SB}} = -1 + [X_\Gamma]_{\mathrm{SB}} = 0$.

It would be interesting to see if there are other natural Rota–Baxter structures on the Grothendieck ring of varieties, or in the ring of immersed conical varieties, that can be used in defining a renormalized version of motivic and algebro-geometric Feynman rules.

Chapter 6

The Riemann–Hilbert correspondence

This chapter is dedicated to the Riemann–Hilbert correspondence of renormalization, as derived in [Connes and Marcolli (2004)]. A very detailed account of this topic is available in [Connes and Marcolli (2008)], hence the treatment given here will be shorter and will only aim at summarizing the main steps involved in the main result of [Connes and Marcolli (2004)]. The interested reader is encouraged to look at [Connes and Marcolli (2006b)] and at the first chapter of [Connes and Marcolli (2008)] for more extensive details, and to [Connes and Marcolli (2005a)] for a shorter survey.

The purpose of this approach is to parameterize the data of divergences in perturbative scalar quantum field theories in terms of suitable gauge equivalence classes of differential systems with irregular singularities. These are then classified in terms of the Riemann–Hilbert correspondence, which means that one shows that these differential systems form a Tannakian category, hence they can be classified in terms of finite dimensional linear representations of an associated affine group scheme. The Riemann–Hilbert correspondence is the equivalence of categories between the analytic data of differential systems and the representation theoretic data.

6.1 From divergences to iterated integrals

The first step in reformulating the divergences of renormalization in terms of differential systems consists of writing the negative piece $\gamma_-(z)$ of the Birkhoff factorization as an iterated integral depending on a single element β in the Lie algebra Lie(G) of the affine group scheme dual to the Connes–Kreimer Hopf algebra. This is a way of formulating what is known in physics as the 't Hooft–Gross relations. These express the fact that counterterms only depend on the beta function of the theory, that is, the infinitesimal generator of the renormalization group flow.

The starting point is the observation that, in the Birkhoff factorization, there is a dependence on a mass (energy) scale μ, inherited from the same dependence of the dimensionally regularized Feynman integrals $U_\mu(\Gamma)$, discussed in (1.55) above. Thus, we have

$$\gamma_\mu(z) = \gamma_-(z)^{-1} \gamma_{\mu,+}(z),$$

where one knows by reasons of dimensional analysis (see [Collins (1986)]) that the negative part is independent of μ.

This negative part $\gamma_-(z)$ can then be written as a time ordered exponential

$$\gamma_-(z) = Te^{-\frac{1}{z} \int_0^\infty \theta_{-t}(\beta) dt} = 1 + \sum_{n=1}^\infty \frac{d_n(\beta)}{z^n},$$

where

$$d_n(\beta) = \int_{s_1 \geq s_2 \geq \cdots \geq s_n \geq 0} \theta_{-s_1}(\beta) \cdots \theta_{-s_n}(\beta) ds_1 \cdots ds_n,$$

and where $\beta \in \mathrm{Lie}(G)$ is the beta function, that is, the infinitesimal generator of renormalization group flow, and the action θ_t is induced by the grading of the Hopf algebra by

$$\theta_u(X) = u^n X, \quad \text{for} \quad u \in \mathbb{G}_m, \quad \text{and} \quad X \in \mathcal{H}, \quad \text{with} \quad \deg(X) = n,$$

with generator the grading operator $Y(X) = nX$. This result follows from the analysis of the renormalization group given in [Connes and Kreimer (2000)] [Connes and Kreimer (2001)], after the recursive formula for the coefficients d_n is explicitly solved, so as to give the time ordered exponential above.

The loop $\gamma_\mu(z)$ collects all the unrenormalized values $U_\mu(\Gamma)$ of the Feynman integrals in the form of an algebra homomorphism $\phi_\mu : \mathcal{H} \to \mathbb{C}(\{z\})$ from the Connes–Kreimer Hopf algebra to the field of germs of meromorphic functions, given on generators by $\phi_\mu(\Gamma) = U_\mu(\Gamma)$.

The renormalization group analysis of [Connes and Kreimer (2001)] shows that $\gamma_\mu(z)$ satisfies the scaling property

$$\gamma_{e^t \mu}(z) = \theta_{tz}(\gamma_\mu(z)) \tag{6.1}$$

in addition to the property that its negative part is independent of μ,

$$\frac{\partial}{\partial \mu} \gamma_-(z) = 0. \tag{6.2}$$

The Birkhoff factorization is then written in [Connes and Marcolli (2004)] in terms of iterated integrals as

$$\gamma_{\mu,+}(z) = T e^{-\frac{1}{z} \int_0^{-z \log \mu} \theta_{-t}(\beta) dt} \, \theta_{z \log \mu}(\gamma_{\text{reg}}(z)).$$

Thus $\gamma_\mu(z)$ is specified by β up to an equivalence given by the regular term $\gamma_{\text{reg}}(z)$. The equivalence corresponds to "having the same negative part of the Birkhoff factorization".

6.2 From iterated integrals to differential systems

Having written the counterterms as iterated integrals has the advantage that it makes it possible to reformulate these data in terms of differential systems. In fact, one can use the fact that iterated integrals are uniquely solutions of certain differential equations.

An iterated integral (or time-ordered exponential) $g(b) = T e^{\int_a^b \alpha(t) dt}$ is the unique solution of a differential equation $dg(t) = g(t)\alpha(t)dt$ with initial condition $g(a) = 1$. In particular, suppose given the differential field $(\mathcal{K}, \delta)$, where $\mathcal{K} = \mathbb{C}(\{z\})$ is the field of convergent Laurent series and $\delta(f) = f'$, and an affine group scheme G. One then has a logarithmic derivative

$$G(\mathcal{K}) \ni f \mapsto D(f) = f^{-1}\delta(f) \in \text{Lie}(G(\mathcal{K})),$$

and one can consider differential equations of the form $D(f) = \omega$, for a flat $\text{Lie}(G(\mathbb{C}))$-valued connection ω, singular at $z = 0 \in \Delta^*$. The existence of solutions is ensured by the condition of trivial monodromy on Δ^*

$$M(\omega)(\ell) = T e^{\int_0^1 \ell^* \omega} = 1, \quad \ell \in \pi_1(\Delta^*).$$

In [Connes and Marcolli (2004)] these differential systems are considered up to a gauge equivalence relation by a regular gauge transformation, $D(fh) = Dh + h^{-1}Df\,h$, for $h \in G(\mathbb{C}\{z\})$, with $\mathbb{C}\{z\}$ the ring of convergent power series (germs of holomorphic functions). This gauge equivalence is the same thing as the requirement that solutions have the same negative piece of the Birkhoff factorization,

$$\omega' = Dh + h^{-1}\omega h \quad \Leftrightarrow \quad f_-^\omega = f_-^{\omega'},$$

where $D(f^\omega) = \omega$ and $D(f^{\omega'}) = \omega'$.

The fact that, by dimensional analysis, counterterms are independent of the energy scale corresponds to the fact that the differential systems are given by connections satisfying an *equisingularity* condition, as we recall below. More precisely, in [Connes and Marcolli (2004)] a geometric reformulation of the two conditions (6.1), (6.2) is given in terms of properties of connections on a principal G-bundle $P = B \times G$ over a fibration $\mathbb{G}_b \to B \to \Delta$, where $z \in \Delta$ is the complexified dimension of DimReg and the fiber $\mu^z \in \mathbb{G}_m$ over z corresponds to the changing mass scale. The multiplicative group acts by

$$u(b, g) = (u(b), u^Y(g)) \quad \forall u \in \mathbb{G}_m.$$

The two conditions (6.1) and (6.2) correspond to the properties that the flat connection ϖ on P^* is *equisingular*, that is, it satisfies:

- Under the action of $u \in \mathbb{G}_m$ the connection transforms like

$$\varpi(z, u(v)) = u^Y(\varpi(z, u)).$$

- If γ is a solution in $G(\mathbb{C}(\{z\}))$ of the equation $D\gamma = \varpi$, then the restrictions along different sections σ_1, σ_2 of B with $\sigma_1(0) = \sigma_2(0)$ have "the same type of singularities", namely

$$\sigma_1^*(\gamma) \sim \sigma_2^*(\gamma),$$

where $f_1 \sim f_2$ means that $f_1^{-1} f_2 \in G(\mathbb{C}\{z\})$, regular at zero.

The first property of equisingular connections corresponds to the scaling property (6.1) for the solutions of the corresponding differential systems, while the second property expresses geometrically the independence of the counterterms from the mass scale, that is, equation (6.2).

6.3   Flat equisingular connections and vector bundles

The setting described above still depends on the choice of a given physical theory through the group G of diffeographisms, dual to the Connes–Kreimer Hopf algebra of the theory.

The affine group scheme G is uniquely determined by its Tannakian category of finite dimensional linear representations Rep_G. Thus, without loss of information, one can replace singular connections on the principal G-bundle $P = B \times G$ with singular connections on associated vector bundles $E = B \times V$ obtained by finite dimensional linear representations of G in $\text{GL}(V)$. This provides a natural larger ambient category where the different diffeographism groups G of different physical theories can be compared. It is obtained by working with all *flat equisingular vector bundles*. Among them one will identify a subcategory of those that are obtained from representations of a given G, thus reconstructing the data associated to divergences of a particular physical theory.

It is proved in [Connes and Marcolli (2004)] (see also Chapter 1 of [Connes and Marcolli (2008)]) that flat equisingular vector bundles form a Tannakian category. We recall here briefly the main steps of the construction of this category.

The objects $\mathrm{Obj}(\mathcal{E})$ are pairs $\Theta = (V, [\nabla])$, where V is a finite dimensional $\mathbb{Z}$-graded vector space, out of which one forms a bundle $E = B \times V$. The vector space has a filtration $W^{-n}(V) = \oplus_{m \geq n} V_m$ induced by the grading and a $\mathbb{G}_m$ action also coming from the grading. The class $[\nabla]$ is an equivalence class of equisingular connections on the vector bundle E, singular over the preimage in B of the point $0 \in \Delta$, which are compatible with the filtration and trivial on the induced graded spaces $Gr^W_{-n}(V)$, up to the equivalence relation of W-equivalence. This is defined by $T \circ \nabla_1 = \nabla_2 \circ T$ for some $T \in \mathrm{Aut}(E)$ which is compatible with the filtration and trivial on $Gr^W_{-n}(V)$. Here the condition that the connections ∇ are equisingular means that they are $\mathbb{G}_m$-invariant and that restrictions of solutions to sections of B with the same $\sigma(0)$ are W-equivalent. The morphisms $\mathrm{Hom}_{\mathcal{E}}(\Theta, \Theta')$ are linear maps $T : V \to V'$ that are compatible with the grading, and such that on $E \oplus E'$ the following connections are W-equivalent:

$$\begin{pmatrix} \nabla' & 0 \\ 0 & \nabla \end{pmatrix} \overset{W-\text{equiv}}{\simeq} \begin{pmatrix} \nabla' & T\nabla - \nabla'T \\ 0 & \nabla \end{pmatrix}. \tag{6.3}$$

Notice that, as usual, the difficult part in defining a good categorical setting is to have the correct notion of morphisms. The reason why (6.3) is the correct notion here is that one needs to bypass a well known obstacle, which also arises when dealing with categories of Hodge structures, namely the fact that *filtered spaces* do not form an abelian category, so the first guess about the choice of morphisms, namely linear maps intertwining the connections and compatible with the filtrations, has to be corrected to account for this problem.

6.4 The "cosmic Galois group"

The proof given in [Connes and Marcolli (2004)] and [Connes and Marcolli (2008)] of the fact that the resulting category $\mathcal{E}$ is Tannakian directly exhibits an equivalence of categories with $\mathrm{Rep}_{\mathbb{U}^*}$ for an affine group scheme $\mathbb{U}^* = \mathbb{U} \rtimes \mathbb{G}_m$ where $\mathbb{U}$ is dual, under the relation $\mathbb{U}(A) = \mathrm{Hom}(\mathcal{H}_{\mathbb{U}}, A)$, to the Hopf algebra $\mathcal{H}_{\mathbb{U}} = U(\mathcal{L})^{\vee}$ dual (as Hopf algebra) to the universal enveloping algebra of the free graded Lie algebra $\mathcal{L} = \mathcal{F}(e_{-1}, e_{-2}, e_{-3}, \cdots)$.

The key step of the proof is showing that, for any given object $\Theta = (V, [\nabla])$ in $\mathcal{E}$ there exists unique $\rho \in Rep_{\mathbb{U}^*}$ such that

$$D\rho(\gamma_{\mathbb{U}}) \overset{W-\mathrm{equiv}}{\simeq} \nabla,$$

where $\gamma_{\mathbb{U}}$ is the *universal singular frame* defined as

$$\gamma_{\mathbb{U}}(z, v) = Te^{-\frac{1}{z} \int_0^v u^Y(e) \frac{du}{u}},$$

where the element $e = \sum_{n=1}^{\infty} e_{-n}$ is the generator of a 1-parameter subgroup of $\mathbb{U}^*$ which lifts the renormalization group at the level of the group $\mathbb{U}^*$. Notice that the infinite sum is well defined since the Lie algebra is pronilpotent.

The group homomorphism $\mathbb{U} \to G$ that realizes the finite dimensional linear representations of G with equisingular connections as a subcategory of $\mathcal{E}$ is obtained by mapping the generators $e_{-n} \mapsto \beta_n$ to the n-th graded piece of the beta function of the theory, seen as an element $\beta = \sum_n \beta_n$ in the Lie algebra $\mathrm{Lie}(G)$.

Thus, the universal singular frame can be regarded as a universal source for the counterterms for all given physical theories, as it maps to the correct counterterms under the map of the generators e_{-n} to the graded components of the beta function.

It is shown in Proposition 1.98 of [Connes and Marcolli (2008)] that the universal singular frame is given, in terms of the generators of the Lie algebra, in the form

$$\gamma_{\mathbb{U}}(-z, v) = \sum_{n \geq 0, k_j > 0} \frac{e_{-k_1} \cdots e_{-k_n}}{k_1(k_1 + k_2) \cdots (k_1 + k_2 + \cdots + k_n)} v^{\sum_j k_j} z^{-n}.$$

The occurrence of these coefficients, which also appear in the local index formula of [Connes and Moscovici (1995)], has a natural explanation in terms of the combinatorics of Dynkin idempotents and Dynkin operators for free Lie algebras, as shown in [Ebrahimi-Fard, Gracia-Bondía, Patras (2007)].

Cartier conjectured the existence of a group, closely related to symmetries of multiple zeta values, that would act on all the coupling constants of scalar quantum field theories, generalizing the renormalization group action. He referred to it as the "cosmic Galois group". The result of [Connes and Marcolli (2004)] reported here above gives a positive answer to Cartier's conjecture. In fact, the Tannakian Galois group $\mathbb{U}^* = \mathbb{U} \rtimes \mathbb{G}_m$ obtained as above acts on the coupling constants via its map to the Connes–Kreimer group of diffeographisms given by the β function, which in turn acts on the coupling constants by the result of [Connes and Kreimer (2000)]. The relation between the affine group scheme $\mathbb{U}^*$ and symmetries of multiple zeta values can be seen via the fact that the same group appears as a motivic Galois group for a category of mixed Tate motives via the result of [Deligne and Goncharov (2005)].

Chapter 7

The geometry of DimReg

Usually, as we have seen in the first chapter of this book, the procedure of dimensional regularization of divergent Feynman integrals, is defined as a purely formal procedure based on analytic continuation to a complex variable of the integral of a Gaussian in dimension D, as a function of D. This means that, while this provides an efficient recipe for analytically continuing Feynman integrals to a complexified dimension, typically no attempt is made to make sense of an underlying geometry in the complexified dimension, that would justify the formal procedures used in dimensional regularization. We describe in this chapter two different possible approaches to a geometric description of dimensional regularization. The first is based on motivic notions and it applies to individual motives (motivic sheaves) associated to given Feynman graphs, in the way we have seen through the previous chapters. This first method was discussed in [Marcolli (2008)]. The second method is based on the use of techniques from noncommutative geometry and was first discussed in Chapter 1 of [Connes and Marcolli (2008)] and in the unpublished manuscript [Connes and Marcolli (2005b)]. We concentrate here especially on the later part of [Connes and Marcolli (2005b)], which was not reported in [Connes and Marcolli (2008)], and which makes contact with the motivic notions considered here, through the language of mixed Hodge structures.

7.1 The motivic geometry of DimReg

Let us consider again the mixed Tate motives (motivic sheaves) defined by the logarithmic extension Log of (2.39) and its symmetric powers Log^n of (2.40), as objects in the triangulated category $\mathcal{DM}_{\mathbb{Q}}(\mathbb{G}_m)$ of motivic sheaves over $\mathbb{G}_m$.

As we have seen, the period matrix (2.48) for the pro-motive Log^∞ of (2.47) defines the mixed Hodge structure associated to the motive. In this Hodge realization, the H^0 piece corresponds to the first column of the matrix M_{Log^n}, where the k-th entry corresponds to the k-th graded piece of the weight filtration.

Let us consider then the corresponding grading operator, that multiplies the k-th entry by T^k. One can then associate to the h^0-piece of the Log^∞ motive the following formal expression that corresponds in the period matrix (2.48) to the H^0 part in the MHS realization:

$$\mathbb{Q} \cdot \sum_k \frac{\log^k(s)}{k!} T^k =: \mathbb{Q} \cdot s^T. \tag{7.1}$$

The formal expression (7.1) has in fact an interpretation in terms of periods. This follows from a well known result (*cf. e.g.* [Goncharov (2001)], Lemma 2.10) expressing the powers of the logarithm in terms of iterated integrals. For iterated integrals we use the notation as in [Goncharov (2001)]

$$\int_a^b \frac{ds}{s} \circ \frac{ds}{s} \circ \cdots \circ \frac{ds}{s} = \int_{a \leq s_1 \leq \cdots \leq s_n \leq b} \frac{ds_1}{s_1} \wedge \cdots \wedge \frac{ds_n}{s_n}. \tag{7.2}$$

We also denote by $\Lambda_{a,b}(n)$ the domain

$$\Lambda_{a,b}(n) = \{(s_1, \ldots, s_n) \mid a \leq s_1 \leq \cdots \leq s_n \leq b\}. \tag{7.3}$$

Lemma 7.1.1. *The expression (7.1) is obtained as rational multiples of the pairing*

$$s^T = \int_{\Lambda_{1,s}(\infty)} \eta(T), \tag{7.4}$$

with $\Lambda_{1,s}(\infty) = \cup_n \Lambda_{1,s}(n)$ and the form

$$\eta(T) := \sum_n \frac{ds_1}{s_1} \wedge \cdots \wedge \frac{ds_n}{s_n} T^n. \tag{7.5}$$

Proof. The result follows from the basic identity (*cf.* [Goncharov (2001)], Lemma 2.10)

$$\int_{\Lambda_{a,b}(n)} \frac{ds_1}{s_1} \wedge \cdots \wedge \frac{ds_n}{s_n} = \frac{\log\left(\frac{b}{a}\right)^n}{n!}. \tag{7.6}$$

$\square$

Observe then that, in terms of the motivic sheaves defined as in [Arapura (2008)] (see the discussion in §2.4 above), one can associate to the graph hypersurfaces X_Γ and the divisors Σ_n with $\partial\sigma_n \subset \Sigma_n$ objects in the Arapura category of motivic sheaves over $\mathbb{G}_m$, by considering the morphism

$$\Psi_\Gamma : \mathbb{A}^{\#E_\Gamma} \smallsetminus \hat{X}_\Gamma \to \mathbb{G}_m \tag{7.7}$$

defined by the graph polynomial $\Psi_\Gamma(s) = \det(M_\Gamma(s))$.

Definition 7.1.2. The category of Feynman motivic sheaves, for a fixed scalar quantum field theory, is the subcategory of the Arapura category of motivic sheaves over $\mathbb{G}_m$ spanned by the objects of the form

$$(\Psi_\Gamma : \mathbb{A}^n \smallsetminus \hat{X}_\Gamma \to \mathbb{G}_m, \hat{\Sigma}_n \smallsetminus (\hat{\Sigma}_n \cap \hat{X}_\Gamma), n - 1, n - 1), \tag{7.8}$$

with $n = \#E_{int}(\Gamma)$ and where Γ ranges over the Feynman graphs of the given scalar field theory, and where $\hat{\Sigma}_n = \{t \in \mathbb{A}^n \mid \prod_i t_i = 0\}$ is the union of the coordinate hyperplanes.

The above correspond to the local systems

$$H^{n-1}_{\mathbb{G}_m}(\mathbb{A}^n \smallsetminus \hat{X}_\Gamma, \hat{\Sigma}_n \smallsetminus (\hat{\Sigma}_n \cap \hat{X}_\Gamma), \mathbb{Q}(n-1)). \tag{7.9}$$

We can then consider the pullback of the logarithmic motive $\mathrm{Log} \in \mathcal{DM}(\mathbb{G}_m)$ by this morphism, as in the construction of the logarithmic specialization system given in [Ayoub (2007)]. This gives a motive

$$\mathrm{Log}_\Gamma := \Psi_\Gamma^*(\mathrm{Log}) \in \mathcal{DM}(U_\Gamma), \tag{7.10}$$

where $U_\Gamma = \mathbb{A}^{\#E_\Gamma} \smallsetminus \hat{X}_\Gamma$.

In terms of periods, for the product of motivic sheaves defined by (2.23) one has the following.

Lemma 7.1.3. *Suppose given $\sigma \subset X$ and $\sigma' \subset X'$, defining relative homology cycles for (X, Σ) and (X', Σ'), respectively. One then has, for the fiber product (2.23), the period pairing*

$$\int_{\sigma \times_S \sigma'} \pi_X^*(\omega) \wedge \pi_{X'}^*(\eta) = \int_\sigma \omega \wedge f^* f'_*(\eta), \tag{7.11}$$

where $f : \sigma \to S$ and $f' : \sigma' \to S$ are the restrictions of the maps $X \to S$ and $X' \to S$.

Proof. First recall that, when integrating a differential form over a fiber product, one has the formula

$$\int_{X \times_S X'} \pi_X^*(\omega) \wedge \pi_{X'}^*(\eta) = \int_X \omega \wedge (\pi_X)_* \pi_{X'}^*(\eta) = \int_X \omega \wedge f^* f'_*(\eta), \tag{7.12}$$

which corresponds to the diagram

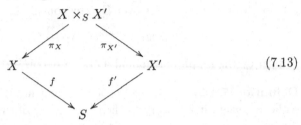

$$(7.13)$$

Suppose then given $\sigma \subset X$ such that $\partial \sigma \subset \Sigma$ and $\sigma' \subset X'$ with $\partial \sigma' \subset \Sigma'$. One has

$$\partial(\sigma \times_S \sigma') = \partial \sigma \times_S \sigma' \cup \sigma \times_S \partial \sigma' \subset \Sigma \times_S X' \cup X \times_S \Sigma',$$

so that $\sigma \times_S \sigma'$ defines a relative homology class in

$$(X \times_S X', \Sigma \times_S X' \cup X \times_S \Sigma').$$

Given elements $[\omega] \in H_{\dot{S}}(X, \Sigma)$ and $[\eta] \in H_{\dot{S}}(X', \Sigma')$, we then apply the formula (7.12) to the integration on $\sigma \times_S \sigma'$ and obtain (7.11). $\square$

In these terms, the procedure of dimensional regularization can then be thought of as follows. Consider again the logarithmic (pro)motive, viewed itself as a motivic sheaf $X_{\mathrm{Log}^\infty} \to \mathbb{G}_m$ over $\mathbb{G}_m$. One can then take the product of a Feynman motive

$$(\Psi_\Gamma : \mathbb{A}^n \smallsetminus \hat{X}_\Gamma \to \mathbb{G}_m, \Sigma_n \smallsetminus (\Sigma_n \cap \hat{X}_\Gamma), n - 1, n - 1), \qquad (7.14)$$

by the (pro)motive

$$(X_{\mathrm{Log}}^\infty \to \mathbb{G}_m, \Lambda_\infty, 0, 0), \qquad (7.15)$$

where Λ_∞ is such that the domain of integration $\Lambda_{1,t}(\infty)$ of the period computation of Lemma 7.1.1 defines a cycle. The product is then given by a fiber product as in (2.23), namely

$$\Psi_\Gamma^*(\mathrm{Log}^\infty) = (\mathbb{A}^n \smallsetminus \hat{X}_\Gamma) \rtimes_{\mathbb{G}_m} X_{\mathrm{Log}^\infty} \longrightarrow X_{\mathrm{Log}^\infty}$$

$$\downarrow \qquad\qquad\qquad\qquad\qquad\qquad \downarrow \qquad\qquad (7.16)$$

$$\mathbb{A}^n \smallsetminus \hat{X}_\Gamma \xrightarrow{\quad \Psi_\Gamma \quad} \mathbb{G}_m$$

We then have the following interpretation of the dimensionally regularized Feynman integrals.

Proposition 7.1.4. *The dimensionally regularized Feynman integrals are periods on the product, in the category of motivic sheaves enlarged to include projective limits, of the Feynman motive (7.14) by the logarithmic pro-motive Log^∞ of (7.15).*

Proof. Consider the product (7.16), with the two projections

$$\pi_X : \Psi_\Gamma^*(\text{Log}^\infty) \to \mathbb{A}^n \smallsetminus \hat{X}_\Gamma$$

$$\pi_L : \Psi_\Gamma^*(\text{Log}^\infty) \to X_{\text{Log}^\infty}$$

and the form $\pi_X^*(\alpha) \wedge \pi_L^*(\eta(T))$, where α is the form involved in the parametric Feynman integral, and $\eta(T)$ is the form on X_{Log^∞} that gives the period (7.4). The period computation of Lemma 7.1.1 gives

$$\Psi_\Gamma^* \left(\int_{\Lambda_{1,s}(\infty)} \eta(T) \right) = \int_{\Lambda_{1,\Psi_\Gamma(t)}(\infty)} \eta(T) = \sum_n \frac{\log(\Psi_\Gamma(t))^n}{n!} T^n = \Psi_\Gamma(t)^T.$$

$$(7.17)$$

We then have, by (7.11),

$$\int_{\sigma_n \times_{\mathbb{G}_m} \Lambda_{1,\Psi_\Gamma(t)}(\infty)} \pi_X^*(\alpha) \wedge \pi_L^*(\eta(T)) = \int_{\sigma_n} \alpha \wedge (\pi_X)_* \pi_L^*(\eta(T)) = \int_{\sigma_n} \Psi_\Gamma^T \alpha.$$

This is the dimensionally regularized integral, up to replacing the formal variable T of (7.1) with the complex DimReg variable z. $\qquad\square$

7.2 The noncommutative geometry of DimReg

We now discuss a different approach to the geometry of DimReg, based on [Connes and Marcolli (2005b)], using spectral triples and noncommutative geometry and we discuss its points of contact with motivic notions.

The notion of metric space in noncommutative geometry is provided by *spectral triples*, see [Connes (1995)]. These consist of data of the form $X = (\mathcal{A}, \mathcal{H}, \mathcal{D})$, with $\mathcal{A}$ an associative involutive algebra represented as an algebra of bounded operators on a Hilbert space $\mathcal{H}$, together with a self-adjoint operator $\mathcal{D}$ on $\mathcal{H}$, with compact resolvent, and with the property that the commutators $[a, \mathcal{D}]$ are bounded operators on $\mathcal{H}$, for all $a \in \mathcal{A}$. A spectral triple is even if there is a linear involution, $\gamma^2 = 1$, on the Hilbert space $\mathcal{H}$ satisfying $[a, \gamma] = 0$ and $D\gamma + \gamma D = 0$. Assume for simplicity that the operator D is invertible. Then the spectral triple is finitely summable if, for some real s, the operator $|D|^s$ is of trace class, $\text{Tr}(|D|^s) < \infty$.

This structure generalizes the data of a compact Riemannian spin manifold, with the (commutative) algebra of smooth functions, the Hilbert space of square integrable spinors and the Dirac operator. It makes sense, however, for a wide range of examples that are not ordinary manifolds, such as quantum groups, fractals, noncommutative tori, etc.

For such spectral triples there are various different notions of dimension. The most sophisticated one is the *dimension spectrum* which is not a single number but a subset of the complex plane consisting of all poles of the family of zeta functions associated to the spectral triple,

$$\text{Dim} = \{ s \in \mathbb{C} | \zeta_a(s) = \text{Tr}(a|D|^{-s}) \text{ have poles } \}.$$

These are points where one has a well defined integration theory on the non-commutative space, the analog of a volume form, given in terms of a residue for the zeta functions.

It is shown in [Connes and Marcolli (2005b)], [Connes and Marcolli (2008)] that there exists a (type II) spectral triple X_z with the properties that the dimension spectrum is $\text{Dim} = \{z\}$ and that one recovers the DimReg prescription for the Gaussian integration in the form

$$\text{Tr}(e^{-\lambda D_z^2}) = \pi^{z/2} \lambda^{-z/2}.$$

The operator D_z is of the form $D_z = \rho(z) F |Z|^{1/z}$, where $Z = F|Z|$ is a self-adjoint operator affiliated to a type II_∞ von Neumann algebra $\mathcal{N}$ and $\rho(z) = \pi^{-1/2}(\Gamma(1 + z/2))^{1/z}$, with the spectral measure $\text{Tr}(\chi_{[a,b]}(Z)) = \frac{1}{2} \int_{[a,b]} dt$, for the type II trace.

The technical aspects of the difference between ordinary spectral triples and type II spectral triples are beyond the purpose of this book, and we simply refer the reader to the relevant literature, especially [Carey, Phillips, Rennie, Sukochev (2006a)] and [Carey, Phillips, Rennie, Sukochev (2006b)].

The ordinary spacetime over which the quantum field theory is constructed can itself be modeled as a (commutative) spectral triple

$$X = (\mathcal{A}, \mathcal{H}, \mathcal{D}) = (\mathcal{C}^\infty(X), L^2(X, S), \slashed{D}_X)$$

and one can take a product $X \times X_z$ given by the cup product of spectral triples (adapted to the type II case)

$$(\mathcal{A}, \mathcal{H}, \mathcal{D}) \cup (\mathcal{A}_z, \mathcal{H}_z, D_z) = (\mathcal{A} \otimes \mathcal{A}_z, \mathcal{H} \otimes \mathcal{H}_z, \mathcal{D} \otimes 1 + \gamma \otimes D_z).$$

This agrees with what is known in physics as the Breitenlohner–Maison prescription to resolve the problem of the compatibility of the chirality operator γ_5 with the DimReg procedure [Breitenlohner and Maison (1977)]. The Breitenlohner–Maison prescription consists of changing the usual Dirac operator to a product, which is indeed of the form as in the cup product of spectral triples,

$$\mathcal{D} \otimes 1 + \gamma \otimes D_z.$$

It is shown in [Connes and Marcolli (2005b)] and [Connes and Marcolli (2008)] that an explicit example of a space X_z that can be used to perform dimensional regularization geometrically can be constructed from the adèle class space, the noncommutative space underlying the spectral realization of the Riemann zeta function in noncommutative geometry (see *e.g.* [Connes (1999)] and Chapters 2 and 4 of [Connes and Marcolli (2008)]), by taking the crossed product of the partially defined action

$$\mathcal{N} = L^\infty(\hat{\mathbb{Z}} \times \mathbb{R}^*) \rtimes \mathrm{GL}_1(\mathbb{Q})$$

and the trace

$$\mathrm{Tr}_\mathcal{N}(f) = \int_{\hat{\mathbb{Z}} \times \mathbb{R}^*} f(1, a) \, da,$$

with the operator

$$Z(1, \rho, \lambda) = \lambda, \quad Z(r, \rho, \lambda) = 0, \ r \neq 1 \in \mathbb{Q}^*.$$

The type II_∞ factor obtained in this way is dual (in the sense of the duality introduced in [Connes (1973)]) to the type III_1 factor associated to the KMS state at "critical temperature" $\beta = 1$ on the Bost–Connes system, see [Bost and Connes (1995)], [Connes and Marcolli (2006a)], and Chapter 3 of [Connes and Marcolli (2008)].

This setting for the geometrization of DimReg provides a way of computing the anomalous graphs of quantum field theory in terms of index cocycles in noncommutative geometry, as explained in [Connes and Marcolli (2005b)] and Chapter 1 of [Connes and Marcolli (2008)]. Here we concentrate on a related aspect that brings us back to motivic notions, namely an interesting formal analogy between the *evanescent gauge potentials* in the noncommutative geometry of DimReg and the complexes of *vanishing and nearby cycles* in algebraic geometry.

Suppose given a spectral triple $(\mathcal{A}, \mathcal{H}, \mathcal{D})$, which may correspond, as above, to the ordinary spacetime over which a given quantum field theory is considered. We assume that it is an even spectral triple, so that one has a chirality operator γ, which is the usual γ_5 in dimension four. Then consider the algebra $\tilde{\mathcal{A}}$ generated by $\mathcal{A}$ and γ, endowed with a $\mathbb{Z}/2\mathbb{Z}$ grading with $\deg(a) = 1$ for $a \in \mathcal{A}$ and $\deg(\gamma) = -1$. The usual commutators are then replaced by graded commutators

$$[\mathcal{D}, a]_- := \mathcal{D}a - (-1)^{\deg(a)} a \mathcal{D} \tag{7.18}$$

We consider then the spectral triple where the Hilbert space and Dirac operator are given by the product with X_z,

$$(\mathcal{H} \otimes \mathcal{H}_z, \mathcal{D} \otimes 1 + \gamma \otimes \mathcal{D}_z)$$

and the algebra is $\tilde{A}$, acting by $a \otimes 1$. We use the notation

$$\mathcal{D}' = \mathcal{D}_z, \quad \bar{\mathcal{D}} = \mathcal{D} \otimes 1, \quad \hat{\mathcal{D}} = \gamma \otimes \mathcal{D}' \quad \mathcal{D}'' = \bar{\mathcal{D}} + \hat{\mathcal{D}}, \qquad (7.19)$$

as in [Connes and Marcolli (2005b)]. The lack of invariance of $\mathcal{D}''$ under the chiral gauge transformation γ gives rise to new evanescent gauge potentials coming from the presence of a nontrivial commutator

$$[\mathcal{D}'', \gamma]_- = 2\gamma\hat{\mathcal{D}}, \qquad (7.20)$$

which disappears when one is back in the case $z = 0$. These evanescent gauge potentials figure prominently in the computations of anomalies in [Connes and Marcolli (2005b)].

We now show that these evanescent gauge potentials also give rise to a complex reminiscent of vanishing cycles. We use an approach based on polarized Hodge–Lefschetz modules, which we have already described in a very similar form in [Consani and Marcolli (2006)] in the context of the archimedean cohomology of [Consani (1998)] and its relation to noncommutative geometry, [Consani and Marcolli (2004)].

The two geometric settings we will compare are the following. On the algebro-geometric side, we consider the case of a geometric degeneration of a family $\mathfrak{X}$ of smooth algebraic varieties over a disk $\Delta \subset \mathbb{C}$. Here $\mathfrak{X}$ is a complex analytic manifold of dimension $\dim_{\mathbb{C}} \mathfrak{X} = n+1$ and $f : \mathfrak{X} \to \Delta$ is a flat, proper morphism with projective fibers. We assume that the map f is smooth on $\mathfrak{X}^* = \mathfrak{X} \smallsetminus Y$ and that Y is a divisor with normal crossings in $\mathfrak{X}$. On the side of noncommutative geometry, we consider the noncommutative spaces $(\mathcal{A}'', \mathcal{H}'', D'')$ obtained by taking the cup product of a spectral triple $(\mathcal{A}, \mathcal{H}, D)$ with a noncommutative space X_z in "complexified dimension" $z \in \Delta$, with the algebra $\mathcal{A}'' = \tilde{A}$ as above.

We shall see that the complex of gauge potentials on $(\mathcal{A}'', \mathcal{H}'', D'')$ behaves in many ways like a complex of forms with logarithmic poles associated to the family $\mathfrak{X}$ of smooth algebraic varieties over the disk $z \in \Delta$. In particular, in the algebro-geometric case, it is known that the special fiber $f^{-1}(0)$ of $f : \mathfrak{X} \to \Delta$ carries a mixed Hodge structure, [Steenbrink (1976)], [Guillén and Navarro Aznar (1990)].

To describe the analogous structure associated to the complex of gauge potentials in the case of a noncommutative space and its deformation to complexified dimension, we take the point of view of Saito's polarized Hodge–Lefschetz modules, [Saito (1988)] and [Guillén and Navarro Aznar (1990)], formulated as in [Consani and Marcolli (2006)].

Let $L = \oplus_{i,j \in \mathbb{Z}} L^{i,j}$ be a finite dimensional bigraded real vector space. Let ℓ_1, ℓ_2 be endomorphisms of L, with $[\ell_1, \ell_2] = 0$ and

$$\ell_1 : L^{i,j} \to L^{i+2,j}, \quad \ell_2 : L^{i,j} \to L^{i,j+2}. \tag{7.21}$$

The data (L, ℓ_1, ℓ_2) define a *bigraded Lefschetz module* if

$$\ell_1^i : L^{-i,j} \to L^{i,j} \quad \text{and} \quad \ell_2^j : L^{i,-j} \to L^{i,j} \tag{7.22}$$

are isomorphisms for $i > 0$ and $j > 0$, respectively.

In the case of a geometric degeneration, with the generic fiber a compact Kähler manifold, one can obtain such a structure from the action of the Lefschetz operator on the primitive part of the cohomology (induced by wedging with the Kähler form) and of the monodromy: the Lefschetz and the log of the monodromy give the endomorphisms ℓ_1, ℓ_2.

Bigraded Lefschetz modules correspond bijectively to finite dimensional representations of $\mathrm{SL}(2, \mathbb{R}) \times \mathrm{SL}(2, \mathbb{R})$. We will use the following notation:

$$\chi(\lambda) := \begin{pmatrix} \lambda & 0 \\ 0 & \lambda^{-1} \end{pmatrix}, \quad u(s) := \begin{pmatrix} 1 & s \\ 0 & 1 \end{pmatrix}, \quad w := \begin{pmatrix} 0 & 1 \\ -1 & 0 \end{pmatrix}, \tag{7.23}$$

for $\lambda \in \mathbb{R}^*$ and $s \in \mathbb{R}$, and

$$u = \begin{pmatrix} 0 & 1 \\ 0 & 0 \end{pmatrix} = \frac{d}{ds} u(s)|_{s=0}.$$

In these terms, the representation $\sigma = \sigma_{(L,\ell_1,\ell_2)}$ satisfies

$$d\sigma(u, 1) = \ell_1, \quad d\sigma(1, u) = \ell_2, \quad \sigma(\chi(\lambda), \chi(t)) = \lambda^i t^j x, \quad \forall x \in L^{i,j}.$$

The data (L, ℓ_1, ℓ_2) define a *bigraded Hodge–Lefschetz module* if all the $L^{i,j}$ have a real Hodge structure and ℓ_1, ℓ_2 are morphisms of Hodge structures.

A *polarization* on (L, ℓ_1, ℓ_2) is a bilinear form

$$\psi : L \otimes L \to \mathbb{R} \tag{7.24}$$

which is compatible with the Hodge structure, satisfies

$$\psi(\ell_k x, y) + \psi(x, \ell_k y) = 0, \quad k = 1, 2, \tag{7.25}$$

and is such that

$$\psi(\cdot, C\ell_1^i \ell_2^j \cdot) \tag{7.26}$$

is symmetric and positive definite on $L^{-i,-j}$. (Here C is the Weil operator.) The data $(L, \ell_1, \ell_2, \psi)$ then define a *bigraded polarized Hodge–Lefschetz module*.

It is also convenient (*e.g.* when working at the level of forms) to consider *differential bigraded polarized Hodge–Lefschetz modules* $(L, \ell_1, \ell_2, \psi, d)$, where

$$d : L^{i,j} \to L^{i+1,j+1}$$

is a differential ($d^2 = 0$) satisfying

$$[\ell_1, d] = 0 = [\ell_2, d]$$

and

$$\psi(dx, y) = \psi(x, dy).$$

In this case, the cohomology $H^*(L, d)$ inherits the structure of a bigraded polarized Hodge–Lefschetz module. Moreover, H^* is identified with $\mathrm{Ker}(\Box)$, where $\Box = d^*d + dd^*$, where

$$d^* = \sigma(w, w)^{-1} \circ d \circ \sigma(w, w),$$

for w as in (7.23) and $\sigma = \sigma_{(L, \ell_1, \ell_2)}$, the representation of $\mathrm{SL}(2, \mathbb{R}) \times \mathrm{SL}(2, \mathbb{R})$.

We now introduce an analog of this structure for noncommutative spaces. We let $(\mathcal{A}, \mathcal{H}, D)$ be an even finitely summable spectral triple. Let γ be the grading operator. We consider the graded algebra $\tilde{\mathcal{A}}$ generated by $\mathcal{A}$ and by γ. Let $(\mathcal{A}'', \mathcal{H}'', D'')$ be the noncommutative space obtained as the product of $(\tilde{\mathcal{A}}, \mathcal{H}, D)$ with the noncommutative space X_z in "complexified dimension" z. Namely, we have

$$\mathcal{A}'' = \tilde{\mathcal{A}}, \qquad \mathcal{H}'' = \mathcal{H} \otimes \mathcal{H}'$$

and

$$D'' = D_z'' = D \otimes 1 + \gamma \otimes D_z',$$

where as above D_z' is the Dirac operator on the space X_z. We let $\Omega_D^*(\mathcal{A})$ denote the complex of gauge potentials of a triple $(\mathcal{A}, \mathcal{H}, D)$, and $H_D^*(\mathcal{A})$ its cohomology.

We begin now by considering the complex $\Omega^{m,r,k}$, with differentials $\delta : \Omega^{m,r,k} \to \Omega^{m+1,r,k}$ and $\delta' : \Omega^{m,r,k} \to \Omega^{m,r+1,k}$. Here we take $\Omega^{m,r,k}$ to be the span of elements of the form

$$\nabla^k(\omega) D^{2r}, \tag{7.27}$$

with $\omega \in \Omega_D^m(\tilde{\mathcal{A}})$. Here we define $\nabla(a) = [D^2, a]$ as before, for an element $a \in \mathcal{A}$ or $a \in [D, \mathcal{A}]$, while we set $\nabla(a\gamma) := [D^2, a]\gamma$, for all $a \in \mathcal{A}$ or $a \in [D, \mathcal{A}]$.

We consider a descending filtration $\cdots \supset F^p \supset F^{p+1} \supset \cdots$ on $\Omega_D^*(\tilde{\mathcal{A}})$ defined by setting

$$F^p \Omega_D^m(\tilde{\mathcal{A}}) = \oplus_{t+s=m,t\geq p} \Omega_D^{t,s}(\tilde{\mathcal{A}}), \tag{7.28}$$

where $\Omega_D^{t,s}(\tilde{\mathcal{A}}) = \Omega_D^t(\mathcal{A})\theta^s$ is the span of $\omega\,\theta^s$, with $\omega \in \Omega_D^t(\mathcal{A})$ and $\theta = \gamma\hat{D}$. We take

$$F^{m+r-k}\Omega_D^m(\tilde{\mathcal{A}}) = \oplus_{t+s=m,s+r\leq k} \Omega_D^t(\mathcal{A})\,\theta^s. \tag{7.29}$$

In this way, we can endow the complex $\Omega^{m,r,k}$ with a tensor product of two "Hodge structures", where the second is 1-dimensional with $F_D^{-r}(\mathbb{C}) := \mathbb{C} \cdot D^{2r}$ and $F_D^{-(r+1)}(\mathbb{C}) = 0$. We have

$$\Omega^{m,r,k} = F^{m+r-k}\Omega_D^m(\tilde{\mathcal{A}}) \otimes_{\mathbb{C}} F_D^{-r}(\mathbb{C}). \tag{7.30}$$

The index of the resulting filtration is $i - k$, where $i = 2r + m$, hence we then take the $\Omega^{m,r,k}$ with the conditions $k \geq 0$ and $k \geq 2r + m$. The differential $d = \delta + \delta'$ on this complex is the same as described above, induced by $da = [D'', a]$ for $a \in \nabla^k(\mathcal{A})$, when decomposing $D'' = D \otimes 1 + \gamma \otimes D'_z$, so that $\delta'a = [\bar{D}, a]$ and $\delta''(a) = [\hat{D}, a]$.

Notice that the decomposition of the total differential $d = \delta + \delta'$, where δ is essentially the original de Rham differential on $\Omega_D^*(\mathcal{A})$ and δ' acts by wedging with the differential $\theta = \gamma\hat{D}$, resembles very closely the case of geometric degenerations, where one also has a total differential $d = \delta + \delta'$, with δ the usual de Rham differential and δ' given by wedging with the form $\theta = f^*(dz/z)$, for $f : \mathfrak{X} \to \Delta$ and z the coordinate on the base Δ (*cf.* [Steenbrink (1976)]).

We introduce endomorphisms $\ell_1 : \Omega^{m,r,k} \to \Omega^{m,r+1,k+1}$ and $\ell_2 : \Omega^{m,r,k} \to \Omega^{m+2,r-1,k}$ of $(\Omega, d = \delta + \delta')$ defined as follows:

$$\ell_1(\nabla^k(\omega)D^{2r}) = \epsilon(\nabla^k(\omega)D^{2r}) = \nabla^{k+1}(\omega)D^{2(r-1)} \tag{7.31}$$

$$\ell_2(\nabla^k(\omega)D^{2r}) = \sqrt{-1}\,\nabla^k(\omega) \wedge \theta^2\,D^{2(r+1)}$$
$$= \sqrt{-1}\,\nabla^k(\omega\,\theta^2)\,D^{2(r+1)}, \tag{7.32}$$

where for simplicity we write θ for the term $\theta = \gamma\hat{D}$.

We then have the following result.

Lemma 7.2.1. *The endomorphisms ℓ_1 and ℓ_2 satisfy $[\ell_1, \ell_2] = 0$ and are compatible with the differential, namely, $[\ell_1, d] = 0$ and $[\ell_2, d] = 0$.*

Proof. It is immediate to verify that $[\ell_1, \ell_2] = 0$. We check compatibility with the differential. We have $[\ell_1, d] = 0$. In fact, we can check the result on elements of the form $\nabla^k(a)D^{2r}$ or $\nabla^k(\gamma a)D^{2r}$ for $a \in \mathcal{A}$. We have

$$(\ell_1 \delta - \delta \ell_1)\nabla^k(a)D^{2r} = [D^2, [D, \nabla^k(a)]]D^{2(r-1)}$$

$$-[D, \nabla^{k+1}(a)]D^{2(r-1)} = 0,$$

since $[D^2, [D, b]] = -[D, [D^2, b]]$, for all $b \in \nabla^k(\mathcal{A})$. We also have

$$(\ell_1 \delta - \delta \ell_1)\nabla^k(\gamma a)D^{2r} = [D^2, [\hat{D}, \gamma \nabla^k(a)]]D^{2(r-1)}$$

$$-[\hat{D}, \gamma \nabla^{k+1}(a)]D^{2(r-1)} = 0,$$

where we use the fact that we have set

$$\nabla(\gamma a) := \gamma \nabla(a)$$

and we get

$$[D^2, [\hat{D}, \gamma b]] = [D^2, 2\gamma b \hat{D}] = 2[D^2, b]\gamma \hat{D}$$

and

$$[\hat{D}, \gamma[D^2, b]] = 2[D^2, b]\gamma \hat{D}.$$

Similarly, we verify that $[\ell_2, d] = 0$. This can be seen easily on elements of the form bD^{2r}, since

$$[D, b(\gamma \hat{D})^2] = -[D, b](\gamma \hat{D})^2,$$

and on elements $\gamma b D^{2r}$, where

$$\ell_2[\hat{D}, \gamma b] = 2\sqrt{-1}\, b\theta^3 = -[\hat{D}, \ell_2(\gamma b)].\qquad \square$$

Thus, ℓ_1, ℓ_2 induce endomorphisms on the cohomology of the double complex.

We introduce an $\mathrm{SL}(2, \mathbb{R}) \times \mathrm{SL}(2, \mathbb{R})$ representation (σ_1, σ_2) on the complex introduced above, which is associated to the operators ℓ_1, ℓ_2.

Definition 7.2.2. Assume that the spectral triple $(\mathcal{A}, \mathcal{H}, D)$ is N-summable with $N = 2n$. Consider σ_1 and σ_2 defined by

$$\sigma_1(\chi(\lambda)) = \lambda^{2r+m}, \qquad \sigma_1(u(s)) = \exp(s\,\ell_1), \qquad \sigma_1(w) = S_1. \qquad (7.33)$$

$$\sigma_2(\chi(\lambda)) = \lambda^{n-m}, \qquad \sigma_2(u(s)) = \exp(s\,\ell_2), \qquad \sigma_2(w) = S_2. \qquad (7.34)$$

Here the operators S_1 and S_2 are defined by powers of ℓ_1, ℓ_2, in the following way. We consider involutions

$$\hat{S}_1 : \Omega^{m,r,k} \to \Omega^{m,-(r+m),k-(2r+m)}$$

$$\hat{S}_2 : \Omega^{m,r,k} \to \Omega^{2n-m,r-(n-m),k}$$

of the form

$$\hat{S}_1(\nabla^k(\omega)D^{2r}) = \ell_1^{-(2r+m)}(\nabla^k(\omega)D^{2r})$$
$$\hat{S}_2(\nabla^k(\omega)D^{2r}) = \ell_2^{n-m}(\nabla^k(\omega)D^{2r}),$$
(7.35)

for $\omega \in \Omega_D^m(\tilde{\mathcal{A}})$. We set $S_1 = \sqrt{-1}^m \hat{S}_1$.

These satisfy the following property.

Proposition 7.2.3. *The data specified in Definition 7.2.2 define a representation of* $\mathrm{SL}(2,\mathbb{R}) \times \mathrm{SL}(2,\mathbb{R})$ *on* (Ω, d).

Proof. In order to show that $\sigma = \sigma_k$, for $k = 1, 2$, defined by (7.33) and (7.34) is indeed a representations of $\mathrm{SL}(2,\mathbb{R})$ it is sufficient ([Lang (1975)] §XI.2) to check that it satisfies the relations

$$\sigma(w)^2 = \sigma(\chi(-1)),$$
$$\sigma(\chi(\lambda))\sigma(u(s))\sigma(\chi(\lambda^{-1})) = \sigma(u(s\lambda^2)).$$
(7.36)

The first relation is clearly satisfied and the second can be verified easily on elements $\nabla^k(\omega)D^{2r}$, with $\omega \in \Omega_D^m(\mathcal{A})$, where we have

$$\sigma_1(\chi(\lambda))\,\sigma_1(u(s))\,\sigma_1(\chi(\lambda^{-1}))\,\nabla^k(\omega)D^{2r} =$$

$$\sigma_1(\chi(\lambda))\left(1 + s\ell_1 + \frac{s^2}{2}\ell_1^2 + \cdots\right)\lambda^{-(2r+m)}\,\nabla^k(\omega)D^{2r} =$$

$$\left(1 + \lambda^{2(r+1)+m}s\ell_1\lambda^{-(2r+m)} + \lambda^{2(r+2)+m}\frac{s^2}{2}\ell_1^2\lambda^{-(2r+m)} + \cdots\right)\nabla^k(\omega)D^{2r}$$

$$= \exp(s\lambda^2\,\ell_1)\,\nabla^k(\omega)D^{2r},$$

and

$$\sigma_2(\chi(\lambda))\,\sigma_2(u(s))\,\sigma_2(\chi(\lambda^{-1}))\,\nabla^k(\omega)D^{2r} =$$

$$\sigma_2(\chi(\lambda))\left(1 + s\ell_2 + \frac{s^2}{2}\ell_2^2 + \cdots\right)\lambda^{n-m}\,\nabla^k(\omega)D^{2r} =$$

$$\left(1 + \lambda^{-n+m+2}s\ell_2\lambda^{n-m} + \lambda^{-n+m+4}\frac{s^2}{2}\ell_2^2\lambda^{n-m} + \cdots\right)\nabla^k(\omega)D^{2r} =$$

$$\exp(s\lambda^2\,\ell_2)\,\nabla^k(\omega)D^{2r}. \qquad \square$$

Notice that, in the case of the representation associated to the "Lefschetz" ℓ_2, the analog of the Hodge $\star$ on forms, which appears in $\sigma_1(w)$, is realized by a power of ℓ_2, by analogy to what happens in the classical case, where the Hodge $\star$ can be realized (on the primitive cohomology) by the $(n - m)$-th power of the Lefschetz operator.

This construction parallels exactly what happens in the construction of the archimedean cohomology of [Consani (1998)] for the fibers at infinity of arithmetic varieties, in the form presented in [Consani and Marcolli (2006)]. This defines a structure that is analogous to the differential bigraded Hodge–Lefschetz modules, by setting $L^{i,j} = \oplus_k \Omega^{m,r,k}$ with $i = 2r + m$ and $j = -n + m$.

We now discuss the polarization. Define the bilinear form ψ by setting

$$\psi(\nabla^k(\omega)D^{2r}, \nabla^{k'}(\omega')D^{2r'}) := \int \gamma \nabla^k(\eta^*) \nabla^{k'}(\eta') D^{2(r+r')} \tag{7.37}$$

where $\omega = \eta\theta^s$ and $\omega' = \eta'\theta^{s'}$.

Lemma 7.2.4. *The bilinear form* (7.37) *satisfies the relation* (7.25).

Proof. For $a = \nabla^k(\eta)\theta^s D^{2r}$ and $b = \nabla^{k'}(\eta')\theta^{s'} D^{2r'}$, we have

$$\psi(\ell_1(a), b) = \int \gamma \nabla^{k+1}(\eta^*) \nabla^{k'}(\eta') \, D^{2(r-1+r')},$$

while

$$\psi(a, \ell_1(b)) = \int \gamma \nabla^k(\eta) \nabla^{k'+1}(\eta') \, D^{2(r+r'-1)}.$$

Thus, integration by parts gives $\psi(\ell_1(a), b) + \psi(a, \ell_1(b)) = 0$.

We also have

$$\psi(\ell_2(a), b) = \int \gamma(-\sqrt{-1}\,\nabla^k(\eta^*))\nabla^{k'}(\eta') \, D^{2(r+1+r')},$$

and

$$\psi(a, \ell_2(b)) = \int \gamma \nabla^k(\eta^*) \sqrt{-1}\,\nabla^{k'}(\eta') \, D^{2(r+r'+1)}.$$

Thus, we also have $\psi(\ell_2(a), b) + \psi(a, \ell_2(b)) = 0$. $\square$

We also have the following result, analogous to the requirement (7.26) for polarizations of Hodge–Lefschetz modules. We work here under an assumption of "tameness" for spectral triples defined in [Várilly and Gracia-Bondía (1993)].

Lemma 7.2.5. *The bilinear form* (7.37) *has the property that*

$$\langle a^*, b \rangle := \psi(a, \ell_1^i \ell_2^j b)$$

agrees on $L^{-i,-j}$ *with the integral with respect to the volume form of* $(\mathcal{A}, \mathcal{H}, D)$. *Under the assumptions of "tameness" for* $(\mathcal{A}, \mathcal{H}, D)$, *it agrees with the inner product on the complex of gauge potentials.*

Proof. For $b = \nabla^{k'}(\eta')\theta^{s'}D^{2r}$ in $L^{-i,-j}$, we have

$$\ell_2^j(b) = \nabla^{k'}(\eta')\theta^{s'+2j}D^{2r+2j}.$$

We then have

$$\ell_1^i \ell_2^j(b) = \nabla^{k'+i}(\eta')\theta^{s'+2j}D^{2r+2j-2i}.$$

This gives

$$\ell_1^i \ell_2^j(b) = \nabla^{k'+2r+m}(\eta')\theta^{s'+2(m-n)}D^{-2r-2n}.$$

Thus, for $a = \nabla^k(\eta)\theta^s D^{2r}$ in $L^{-i,-j}$, we obtain

$$\psi(a, \ell_1^i \ell_2^j(b)) = \int \gamma \nabla^k(\eta^*)\nabla^{k'+2r+m}(\eta')D^{-2n}.$$

Recall that $\int a := \int a D^{-2n}$ is the integration with respect to the volume form D^{-2n} of the spectral triple $(\mathcal{A}, \mathcal{H}, D)$. Under the assumption of "tameness" this satisfies $\int ab = \int ba$ and $\int a^*a \geq 0$, so that $\langle a^*, b \rangle$ agrees with the inner product of forms in $\Omega_D^*(\tilde{\mathcal{A}})$. $\square$

The fact that the structure described here in terms of Hodge–Lefschetz modules parallels very closely the construction of the archimedean cohomology of [Consani (1998)], in the form presented in [Consani and Marcolli (2006)], suggests that the formalism of DimReg via noncommutative geometry may be useful also to describe a "neighborhood" of the fibers at infinity of an arithmetic variety.

In the algebro-geometric setting of a degeneration over a disk, the local monodromy plays an important role in determining the limiting mixed Hodge structure. In fact, geometrically, the difference between the cohomology of the generic fiber and of the special fiber is measured by the *vanishing cycles*. These span the reduced cohomology $\tilde{H}^*(M_z, \mathbb{C})$ of the Milnor fiber $M_z := B_P \cap f^{-1}(z)$, defined for $P \in Y$, B_P a small ball around P, and a sufficiently small $z \in \Delta^*$. The *nearby cycles* span the cohomology $H^*(M_z, \mathbb{C})$. To eliminate the non-canonical dependence of everything upon the choice of z, one usually considers all choices by passing to the universal cover $\tilde{\Delta}^* = \mathbb{H}$ of the punctured disk and replacing M_z by $\tilde{\mathfrak{X}}^* = \mathfrak{X} \times_\Delta \tilde{\Delta}^*$. The identification (*cf.* [Steenbrink (1976)])

$$H^m(\tilde{\mathfrak{X}}^*, \mathbb{C}) \xrightarrow{\cong} \mathbb{H}^m(Y, \Omega^{\cdot}_{\mathfrak{X}/\Delta}(\log Y) \otimes_{\mathcal{O}_{\mathfrak{X}}} \mathcal{O}_Y) \qquad (7.38)$$

shows that, when working with nearby cycles, one can use a complex of forms with logarithmic differentials. This approach, with an explicit resolution of the complex, was used ([Steenbrink (1976)], [Guillén and Navarro Aznar (1990)]) to obtain a mixed Hodge structure on $H^m(\tilde{\mathfrak{X}}^*, \mathbb{C})$ determined by $(H^m(\tilde{\mathfrak{X}}^*, \mathbb{C}), L_\cdot, F^\cdot)$, where $F^\cdot$ is the Hodge filtration and $L_\cdot$ is the Picard–Lefschetz filtration associated to the local monodromy.

The analogy described above with the structure of polarized Hodge–Lefschetz modules allows one to think of the operator

$$\Theta(a) = \sum_{n=1}^{\infty} \frac{(-1)^{n+1}}{n} \nabla^n(a) D^{-2n}. \qquad (7.39)$$

as the *logarithm of the local monodromy*, with ℓ_1 satisfying $e^\Theta = 1 + \ell_1$, and with the induced action of $(2\pi\sqrt{-1})\ell_1$ on the cohomology corresponding to the residue of the Gauss–Manin connection in the algebro-geometric setting.

The operator Θ of (7.39) is the derivation that appears in [Connes and Moscovici (1995)] in the context of the local index formula for spectral triples.

Chapter 8

Renormalization, singularities, and Hodge structures

As we have seen, the graph hypersurfaces X_Γ are typically singular with non-isolated singularities and a singular locus of small codimension. Among various techniques introduced for the study of non-isolated singularities, a common procedure consists of cutting the ambient space with linear spaces of dimension complementary to that of the singular locus of the hypersurface (*cf. e.g.* [Teissier (1975)]). In this case, the restriction of the function defining the hypersurface to these linear spaces defines hypersurfaces with isolated singularities, to which the usual invariants and constructions for isolated singularities can be applied.

In many cases one ends up slicing the graph hypersurfaces by planes $\mathbb{P}^2$ intersecting the hypersurface into a curve with isolated singular points. This corresponds to cases where the singular locus is of (complex) codimension one in the hypersurface. When the singular locus is of codimension two in the hypersurface, the slicing is given by 3-dimensional spaces cutting the hypersurface into a family of surfaces in $\mathbb{P}^3$ with isolated singularities.

The parametric Feynman integrals can in turn be expressed in terms of this slicing, using projective Radon transforms, and we use the resulting integrals on the slices as a way to relate the Feynman integrals to Hodge structures for isolated singularities. We follow here the approach outlined in [Marcolli (2008)].

8.1 Projective Radon transform

We recall the basic setting for integral transforms on projective spaces (*cf.* §II of [Gelfand, Gindikin, and Graev (1980)]). On any k-dimensional subspace $\mathbb{A}^k \subset \mathbb{A}^n$ there is a unique (up to a multiplicative constant) $(k-1)$-form that is invariant under the action of SL_k. It is given as in (3.42) by the expression

$$\Omega_k = \sum_{i=1}^{k} (-1)^{i+1} t_i \, dt_1 \wedge \cdots \wedge \widehat{dt_i} \wedge \cdots \wedge dt_k. \tag{8.1}$$

The form (8.1) is homogeneous of degree k. Suppose given a function f on $\mathbb{A}^n$ which satisfies the homogeneity condition

$$f(\lambda t) = \lambda^{-k} f(t), \quad \forall t \in \mathbb{A}^n, \lambda \in \mathbb{G}_m. \tag{8.2}$$

Then the integrand $f \Omega_k$ is well defined on the corresponding projective space $\mathbb{P}^{k-1} \subset \mathbb{P}^{n-1}$ and one defines the integral by integrating on a fundamental domain in $\mathbb{A}^k \smallsetminus \{0\}$, *i.e.* on a surface that intersects each line from the origin once.

Suppose given linearly independent dual vectors $\xi_i \in (\mathbb{A}^n)'$, for $i = 1, \ldots, n - k$. These define a k-dimensional linear subspace $\Pi = \Pi_\xi \subset \mathbb{A}^n$ by the vanishing

$$\Pi_\xi = \{ t \in \mathbb{A}^n \mid \langle \xi_i, t \rangle = 0, \, i = 1, \ldots, n - k \}. \tag{8.3}$$

Given a choice of a subspace Π_ξ, there exists a $(k-1)$-form Ω_ξ on $\mathbb{A}^n$ satisfying

$$\langle \xi_1, dt \rangle \wedge \cdots \wedge \langle \xi_{n-k}, dt \rangle \wedge \Omega_\xi = \Omega_n, \tag{8.4}$$

with Ω_n the $(n-1)$-form of (3.42), *cf.* (8.1). The form Ω_ξ is not uniquely defined on $\mathbb{A}^n$, but its restriction to Π_ξ is uniquely defined by (8.4). Then, given a function f on $\mathbb{A}^n$ with the homogeneity condition (8.2), one can consider the integrand $f \Omega_\xi$ and define its integral over the projective space $\pi(\Pi_\xi) \subset \mathbb{P}^{n-1}$ as above. This defines the integral transform, that is, the $(k-1)$-dimensional projective Radon transform (§II of [Gelfand, Gindikin, and Graev (1980)]) as

$$\mathcal{F}_k(f)(\xi) = \int_{\pi(\Pi_\xi)} f(t) \, \Omega_\xi(t) = \int_{\mathbb{P}^{n-1}} f(t) \prod_{i=1}^{n-k} \delta(\langle \xi_i, t \rangle) \, \Omega_\xi(t). \tag{8.5}$$

For our purposes, it is convenient to also consider the following variant of the Radon transform (8.5).

Definition 8.1.1. Let $\sigma \subset \mathbb{A}^n$ be a compact region that is contained in a fundamental domain of the action of $\mathbb{G}_m$ on $\mathbb{A}^n \smallsetminus \{0\}$. The partial $(k-1)$-dimensional projective Radon transform is given by the expression

$$\mathcal{F}_{\sigma, k}(f)(\xi) = \int_{\sigma \cap \pi(\Pi_\xi)} f(t) \, \Omega_\xi(t) = \int_{\sigma \cap \pi(\Pi_\xi)} f(t) \prod_{i=1}^{n-k} \delta(\langle \xi_i, t \rangle) \, \Omega_\xi(t), \tag{8.6}$$

where one identifies σ with its image $\pi(\sigma) \subset \mathbb{P}^{n-1}$.

Let us now return to the parametric Feynman integrals we are considering.

Proposition 8.1.2. *The parametric Feynman integral can be reformulated as*

$$U(\Gamma) = \frac{\Gamma(k - \frac{D\ell}{2})}{(4\pi)^{\ell D/2}} \int \mathcal{F}_{\sigma_n,k}(f_\Gamma)(\xi) \langle \xi, dt \rangle, \tag{8.7}$$

where ξ is an $(n-k)$-frame in $\mathbb{A}^n$ and $\mathcal{F}_{\sigma_n,k}(f)$ is the Radon transform, with σ_n the simplex $\sum_i t_i = 1$, $t_i \geq 0$, and with

$$f_\Gamma(t) = \frac{V_\Gamma(t,p)^{-k+D\ell/2}}{\Psi_\Gamma(t)^{D/2}}. \tag{8.8}$$

Proof. We write the parametric Feynman integral as

$$U(\Gamma) = (4\pi)^{-\ell D/2} \int_{\mathbb{A}^n} \chi_+(t) \frac{e^{-V_\Gamma(t,p)}}{\Psi_\Gamma(t)^{D/2}} \, dt_1 \cdots dt_n, \tag{8.9}$$

where $\chi_+(t)$ is the characteristic function of the domain $\mathbb{R}_+^n$.

Given a choice of an $(n-k)$-frame ξ, we can then write the Feynman integrals in the form

$$U(\Gamma) = (4\pi)^{-\ell D/2} \int \left(\int_{\Pi_\xi} \chi_+(t) \frac{e^{-V_\Gamma(t,p)}}{\Psi_\Gamma(t)^{D/2}} \omega_\xi \right) \langle \xi, dt \rangle, \tag{8.10}$$

where $\langle \xi, dt \rangle$ is a shorthand notation for

$$\langle \xi, dt \rangle = \langle \xi_1, dt \rangle \wedge \cdots \wedge \langle \xi_{n-k}, dt \rangle$$

and ω_ξ satisfies

$$\langle \xi, dt \rangle \wedge \omega_\xi = \omega_n = dt_1 \wedge \cdots \wedge dt_n. \tag{8.11}$$

We then follow the same procedure we used in the derivation of the parametric form of the Feynman integral, applied to the integral over Π_ξ and we write it in the form

$$\int_{\Pi_\xi} \chi_+(t) \frac{e^{-V_\Gamma(t,p)}}{\Psi_\Gamma(t)^{D/2}} \omega_\xi(t) =$$
$$\Gamma(k - \frac{D\ell}{2}) \int_{\Pi_\xi} \delta(1 - \sum_i t_i) \frac{\omega_\xi(t)}{\Psi_\Gamma(t)^{D/2} V_\Gamma(t,p)^{k-D\ell/2}}. \tag{8.12}$$

The function $f_\Gamma(t)$ of (8.8) satisfies the scaling property (8.2) and the integrand

$$\frac{\omega_\xi(t)}{\Psi_\Gamma(t)^{D/2} V_\Gamma(t,p)^{k-D\ell/2}}$$

is therefore $\mathbb{G}_m$-invariant, since the form ω_ξ is homogeneous of degree k. Moreover, the domain σ_n of integration is contained in a fundamental domain for the action of $\mathbb{G}_m$. Thus, we can reformulate the integral (8.12) in projective space, in terms of the Radon transform, as

$$\Gamma(k - \frac{D\ell}{2})\,(4\pi)^{-\ell D/2} \int \mathcal{F}_{\sigma_n,k}(f_\Gamma)(\xi)\,\langle \xi, dt \rangle, \tag{8.13}$$

where $\mathcal{F}_{\sigma_n,k}(f_\Gamma)$ is the Radon transform over the simplex σ_n, as in Definition 8.1.1. $\qquad\Box$

In the following, we will then consider integrals of the form

$$\mathbb{U}(\Gamma)_\xi = \mathcal{F}_{\sigma_n,k}(f_\Gamma)(\xi) = \int_{\Pi_\xi} \delta(1 - \sum_i t_i)\, \frac{\omega_\xi(t)}{\Psi_\Gamma(t)^{D/2} V_\Gamma(t,p)^{k-D\ell/2}} \tag{8.14}$$

$$= \int_{\sigma_\xi} \frac{\omega_\xi(t)}{\Psi_\Gamma(t)^{D/2} V_\Gamma(t,p)^{k-D\ell/2}}$$

as well as their dimensional regularizations

$$\mathbb{U}(\Gamma)_\xi(z) = \int_{\sigma_\xi} \frac{\omega_\xi(t)}{\Psi_\Gamma(t)^{(D+z)/2} V_\Gamma(t,p)^{k-(D+z)\ell/2}}, \tag{8.15}$$

where Π_ξ is a generic linear subspace of dimension equal to the codimension of the singular locus of the hypersurface $X_\Gamma \cup Y_\Gamma$, with $X_\Gamma = \{\Psi_\Gamma(t) = 0\}$ and $Y_\Gamma = \{P_\Gamma(t,p) = 0\}$ as before, and $\sigma_\xi = \sigma_n \cap \Pi_\xi$.

So far we have always made the simplifying assumption, in dealing with the parametric Feynman integrals, that the spacetime dimension D is in the "stable range" where $n \leq D\ell/2$, so that the integral lives on the hypersurface complement of the graph hypersurface X_Γ and the zeros of the second graph polynomial $P_\Gamma(t,p)$ do not contribute to singularities of the differential form. Here the corresponding condition would then be $k \leq D\ell/2$, which is also satisfied in the original stable range, so we continue to restrict only to that case.

8.2 The polar filtration and the Milnor fiber

Algebraic differential forms $\omega \in \Omega^k(\mathcal{D}(f))$ on a hypersurface complement can always be written in the form $\omega = \eta/f^m$ as in (3.46), for some $m \in \mathbb{N}$ and some $\eta \in \Omega^k_{m\,\deg(f)}$. The minimal m such that ω can be written in the form $\omega = \eta/f^m$ is called the order of the pole of ω along the hypersurface X and is denoted by $\mathrm{ord}_X(\omega)$. The order of the pole induces a filtration, called the *polar filtration*, on the de Rham complex of differential forms on the hypersurface complement. One denotes by $P^r\Omega^k_{\mathbb{P}^n} \subset \Omega^k_{\mathbb{P}^n}$ the subspace of forms of order $\mathrm{ord}_X(\omega) \leq k - r + 1$, if $k - r + 1 \geq 0$, or $P^r\Omega^k = 0$ for $k - r + 1 < 0$. The polar filtration $P^\bullet$ is related to the Hodge filtration $F^\bullet$ by $P^r\Omega^m \supset F^r\Omega^m$, by a result of [Deligne and Dimca (1990)].

Proposition 8.2.1. *Under the generic condition on the external momenta, and in the stable range $n \leq D\ell/2$, the forms*

$$\frac{\Omega_\xi}{\Psi_\Gamma^{D/2} V_\Gamma^{k-D\ell/2}} \tag{8.16}$$

span subspaces $P^{r,k}_\xi$ of the polar filtration $P^r\Omega^{k-1}_{\mathbb{P}^{n-1}}$ of a hypersurface complement $\mathcal{U}(f) \subset \mathbb{P}^{n-1}$, where $f = \Psi_\Gamma$ and for the index r of the filtration in the range $r \leq 2k - D(\ell+1)/2$.

Proof. We are assuming that P_Γ and Ψ_Γ have no common factor, for generic external momenta. We also have that $k \leq D\ell/2$, so that also $k - D(\ell+1)/2 < 0$. In this case, we write (8.16) in the form

$$\frac{\Delta(\alpha)}{f^m} = \frac{P_\Gamma^{-k+D\ell/2}\Omega_\xi}{\Psi_\Gamma^{-k+D(\ell+1)/2}}, \tag{8.17}$$

where

$$\alpha = P_\Gamma^{-k+D\ell/2}\omega_\xi, \quad \text{and} \quad f = \Psi_\Gamma \quad \text{and} \quad m = -k + D(\ell+1)/2. \tag{8.18}$$

We are considering here the polar filtration on forms on the complement of the hypersurface X_Γ defined by $\Psi_\Gamma = 0$. Varying the external momenta p correspondingly varies the polynomial $P_\Gamma(t, p)$. We then obtain, for generic ξ and varying p, a subspace of $P^{r,k}_\xi$ of the filtration $P^r\Omega^{k-1}_{\mathbb{P}^{n-1}}$, for all $r \leq 2k - D(\ell+1)/2$. □

Suppose then that $k = \mathrm{codim}\,\mathrm{Sing}(X)$, where $\mathrm{Sing}(X)$ is the singular locus of the hypersurface $X = \{f = 0\}$, with f as in Proposition 8.2.1 above. In this case, for generic ξ, the linear space Π_ξ cuts the singular locus $\mathrm{Sing}(X)$ transversely and the restriction $X_\xi = X \cap \Pi_\xi$ has isolated singularities.

Recall that, in the case of isolated singularities, there is an isomorphism between the cohomology of the Milnor fiber F_ξ of X_ξ and the total cohomology of the Koszul–de Rham complex of forms (3.46) with the total differential $d_f\omega = f\,d\omega - m\,df \wedge \omega$ as in (3.51).

The explict isomorphism is given by the Poincaré residue map and can be written in the form

$$[\omega] \mapsto [j^*\Delta(\omega_\xi)], \tag{8.19}$$

where $j : F_\xi \hookrightarrow \Pi_\xi$ is the inclusion of the Milnor fiber in the ambient space (see [Dimca (1992)], §6).

Let $M(f)$ be the Milnor algebra of f, *i.e.* the quotient of the polynomial ring in the coordinates of the ambient projective space by the ideal of the derivatives of f. When f has isolated singularities, the Milnor algebra is finite dimensional. One denotes by $M(f)_m$ the homogeneous component of degree m of $M(f)$.

It then follows from the identification (8.19) above ([Dimca (1992)],§6.2) that, in the case of isolated singularities, a basis for the cohomology $H^r(F_\xi)$ of the Milnor fiber, with $r = \dim \Pi_\xi - 1$ is given by elements of the form

$$\omega_\alpha = \frac{t^\alpha \Delta(\omega_\xi)}{f^m}, \qquad \text{with} \qquad t^\alpha \in M(f)_{m \deg(f) - k}, \tag{8.20}$$

where f is the restriction to Π_ξ of the function of Ψ_Γ. We then have the following consequence of Proposition 8.2.1.

Corollary 8.2.2. *For $n \le D\ell/2$ and for a generic $(n-k)$-frame ξ with $n - k = \dim \mathrm{Sing}(X)$, with $X = X_\Gamma$, and for a fixed generic choice of the external momenta p under the generic assumption of Definition 3.3.1, the Feynman integrand (8.16) of (8.14) defines a cohomology class in $H^r(F_\xi)$, with $r = \dim \Pi_\xi - 1$ and $F_\xi \subset \Pi_\xi$ the Milnor fiber of the hypersurface with isolated singularities $X_\xi = X \cap \Pi_\xi \subset \Pi_\xi$.*

Proof. By Proposition 8.2.1, in the stable range for D, the form (8.16) can be written as

$$\frac{h \Delta(\omega_\xi)}{f^m}, \tag{8.21}$$

where $f = \Psi_\Gamma$ and $h = P_\Gamma^{-k+D\ell/2}$. Let $\mathcal{I}_\xi$ denote the ideal of derivatives of the restriction $f|_{\Pi_\xi}$ of f to Π_ξ. Then let

$$h_\xi = h \mod \mathcal{I}_\xi. \tag{8.22}$$

For a fixed generic choice of the external momenta, this defines an element in the Milnor algebra $M(f|_{\Pi_\xi})$, which lies in the homogeneous component $M(f|_{\Pi_\xi})_{m \deg(f) - k}$, for $m = -k + D(\ell+1)/2$. Thus, the form (8.21) defines a class in the cohomology $H^r(F_\xi)$ with $r = \dim \Pi_\xi - 1$. $\qquad\square$

We have not discussed so far the dependence of our slicing method upon the choice of the slice. This will be treated in §8.4 below, where we show that one can include the choice of slices as additional data in the Hopf algebra combinatorics, so as to consider all slices simultaneously, with a combinatorial organizing principle. In particular, notice that in terms of the Radon transform decomposition (8.10) only the general slices matter as the special slices (for instance those where Π_ξ is contained in the hypersurface X_Γ) form a set of measure zero in the space of $(n - k)$-frames.

It is also worth pointing out that the construction described in this section can also be formulated without slicing. Namely, the operations of taking the ideal of derivatives of f, finding the class of the numerator modulo these derivatives, and then ranging over all choices of external momenta to get a subspace can be performed without the slicing operation. What is missing in this case is the interpretation of the resulting space as the cohomology of a Milnor fiber.

8.3 DimReg and mixed Hodge structures

We assume here to be in the case of isolated singularities, possibly after replacing the original Feynman integrals with their slices along planes Π_ξ of dimension complementary to that of the singular locus of the hypersurface, as above.

The cohomological Milnor fibration has fiber over ϵ given by the complex vector space $H^{k-1}(F_\epsilon, \mathbb{C})$, where the Milnor fiber F_ϵ of X_ξ is homotopically a bouquet of μ spheres S^{k-1}, with $k = \dim \Pi_\xi - 1$ and with μ the Milnor number of the isolated singularity. A holomorphic k-form $\alpha = h\omega_\xi/f^m$ determines a section of the cohomological Milnor fibration by taking the classes

$$\left[\frac{\alpha}{df}|_{F_\epsilon} \right] \in H^{k-1}(F_\epsilon, \mathbb{C}). \tag{8.23}$$

We then have the following results ([Arnold, Gusein-Zade, Varchenko (1988)], Vol.II §13). The asymptotic formula for the Gelfand–Leray functions implies that the function of ϵ obtained by pairing the section (8.23) with a locally constant section of the homological Milnor fibration has an asymptotic expansion

$$\left\langle \left[\frac{\alpha}{df} \right], \delta \right\rangle \sim \sum_{\lambda, r} \frac{a_{r,\lambda}}{r!} \epsilon^\lambda \log(\epsilon)^r, \tag{8.24}$$

for $\epsilon \to 0$, where $\delta(\epsilon) \in H_{k-1}(F_\epsilon, \mathbb{Z})$. Moreover, there exist classes

$$\eta^\alpha_{r,\lambda}(\epsilon) \in H^{k-1}(F_\epsilon, \mathbb{C}) \tag{8.25}$$

such that the coefficients $a_{r,\lambda}$ of (8.24) are given by

$$\langle \eta^\alpha_{r,\lambda}(\epsilon), \delta(\epsilon) \rangle = a_{r,\lambda}. \tag{8.26}$$

Thus, one defines the "geometric section" associated to the holomorphic k-form α as

$$\sigma(\alpha) := \sum_{r,\lambda} \eta_{r,\lambda}^{\alpha}(\epsilon) \frac{\epsilon^{\lambda} \log(\epsilon)^r}{r!}. \qquad (8.27)$$

The order of the geometric section $\sigma(\alpha)$ is defined as being the smallest λ in the discrete set $\Xi \subset \mathbb{R}$ such that $\eta_{0,\lambda}^{\alpha} \neq 0$. One denotes it by λ_{α}. The *principal part* of $\sigma(\alpha)$ is then defined as

$$\sigma_{\max}(\alpha)(\epsilon) := \epsilon^{\lambda_{\alpha}} \left(\eta_{0,\lambda_{\alpha}}^{\alpha} + \cdots + \frac{\log(\epsilon)^{k-1}}{(k-1)!} \eta_{k-1,\lambda_{\alpha}}^{\alpha} \right), \qquad (8.28)$$

where one knows that

$$\eta_{r,\lambda}^{\alpha} = \mathcal{N}^r \eta_{0,\lambda}^{\alpha}, \qquad (8.29)$$

where $\mathcal{N}$ is the nilpotent operator given by the logarithm of the unipotent monodromy, given by

$$\mathcal{N} = -\frac{1}{2\pi i} \log \mathcal{T}$$

with $\log \mathcal{T} = \sum_{r \geq 1} (-1)^{r+1} (\mathcal{T} - id)^r / r$.

The *asymptotic mixed Hodge structure* on the fibers of the cohomological Milnor fibration constructed by Varchenko ([Varchenko (1980)], [Varchenko (1981)]) has as the Hodge filtration the subspaces $F^r \subset H^{k-1}(F_{\epsilon}, \mathbb{C})$ defined by

$$F^r = \{[\alpha/df] \,|\, \lambda_{\alpha} \leq k - r - 1\} \qquad (8.30)$$

and as weight filtration $W_{\ell} \subset H^{k-1}(F_{\epsilon}, \mathbb{C})$ the filtration associated to the nilpotent monodromy operator $\mathcal{N}$. This mixed Hodge structure has the same weight filtration as the *limiting mixed Hodge structure* constructed by Steenbrink ([Steenbrink (1976)], [Steenbrink (1977)]), but the Hodge filtration is different, though the two agree on the graded pieces of the weight filtration.

We show that, upon varying the choice of the external momenta p and of the spacetime dimension D, the corresponding Feynman integrands, in a neighborhood of an isolated singular point of $X_\Gamma \cap \Pi_\xi$, determine a subspace of the cohomology $H^{k-1}(F_\xi, \mathbb{C})$ of the Milnor fiber of $X_\Gamma \cap \Pi_\xi$. This inherits a Hodge and a weight filtration from the Milnor fiber cohomology with its asymptotic mixed Hodge structure. We concentrate again only on the case where $k - D\ell/2 \leq 0$, so that we can consider, for fixed k, arbitrarily large values of $D \in \mathbb{N}$.

Proposition 8.3.1. *Consider Feynman integrals, sliced along a linear space* Π_ξ *as in (8.14). We write the integrand in the form*

$$\alpha_\xi = \frac{h\Omega_\xi}{f^m}, \tag{8.31}$$

with

$$\begin{cases} h = P_\Gamma^{-k+D\ell/2} \\ f = \Psi_\Gamma \\ m = -k + D(\ell+1)/2, \end{cases} \tag{8.32}$$

as in (8.18), with $k - D\ell/2 \leq 0$. *Upon varying the external momenta* p *in* $P_\Gamma(p,t)$ *and the spacetime dimension* $D \in \mathbb{N}$, *with* $k - D\ell/2 \leq 0$, *the forms* α_ξ *as above determine a subspace*

$$H_{\text{Feynman}}^{k-1}(F_\epsilon, \mathbb{C}) \subset H^{k-1}(F_\epsilon, \mathbb{C}),$$

of the fibers of the cohomological Milnor fibration, spanned by elements of the form (8.31), where the polynomials $h = h_{T,v,w,p}$ *are of the form*

$$h(t) = \prod_{i=1}^{-k+D\ell/2} L_{T_i}(t) \prod_{e \notin T_i} t_e, \tag{8.33}$$

where the T_i *are spanning trees and the* $L_{T_i}(t)$ *are the linear functions of (3.41).*

Proof. Consider the explicit expression of the polynomial $P_\Gamma(t,p)$ as a function of the external momenta, through the coefficients s_C of (3.18). One can see that, by varying arbitrarily the external momenta, subject to the global conservation law (3.39), one can reduce to the simplest possible case, where all external momenta are zero except for a pair of opposite momenta $P_{v_1} = p = -P_{v_2}$ associated to a pair of external edges attached to a pair of vertices v_1, v_2. In such a case, the polynomial $P_\Gamma(t,p)$ becomes of the form (3.40). Thus, when considering powers $P_\Gamma(t,p)^{-k+D\ell/2}$ for varying D, we obtain all polynomials of the form (8.33). $\qquad\square$

We denote by $H_{\text{Feynman}}^{k-1}(F_\epsilon, \mathbb{C})$ the subspace of the cohomology $H^{k-1}(F_\epsilon, \mathbb{C})$ of the Milnor fiber spanned by the classes $[\alpha_\xi/df]$ with α_ξ of the form (8.31), with h of the form (8.33), considered modulo the ideal generated by the derivatives of $f = \Psi_\Gamma$ and localized at an isolated singular point, *i.e.* viewed as elements in the Milnor algebra $M(f)$. The subspace $H_{\text{Feynman}}^{k-1}(F_\epsilon, \mathbb{C})$ inherits a Hodge and a weight filtration

$F^\bullet \cap H^{k-1}_{\text{Feynman}}$ and $W_\bullet \cap H^{k-1}_{\text{Feynman}}$ from the asymptotic mixed Hodge structure of Varchenko on $H^{k-1}(F_\epsilon, \mathbb{C})$. It is an interesting problem to see whether the subspace H^{k-1}_{Feynman} recovers the full $H^{k-1}(F_\epsilon, \mathbb{C})$ and if $(F^\bullet \cap H^{k-1}_{\text{Feynman}}, W_\bullet \cap H^{k-1}_{\text{Feynman}})$ still give a mixed Hodge structure, at least for some classes of graphs Γ.

8.4 Regular and irregular singular connections

An important and still mysterious aspect of the motivic approach to Feynman integrals and renormalization is the problem of reconciling the Riemann–Hilbert correspondence of perturbative renormalization formulated in [Connes and Marcolli (2004)] (see Chapter 1 of [Connes and Marcolli (2008)]), based on equivalence classes of certain *irregular singular* connections, with the setting of motives (especially mixed Tate motives) and mixed Hodge structures, which are naturally related to *regular singular* connections. The irregular singular connections of [Connes and Marcolli (2004)] have values in the Lie algebra of the Connes–Kreimer group of diffeographisms and are defined on a fibration over a punctured disk with fiber the multiplicative group, respectively representing the complex variable z of dimensional regularization and the energy scale μ (or rather μ^z) upon which the dimensionally regularized Feynman integrals depend. On the other hand, in the case of hypersurfaces in projective spaces, the natural associated regular singular connection is the Gauss–Manin connection on the cohomology of the Milnor fiber and the Picard–Fuchs equation for the vanishing cycles. We sketch here a relation between this regular singular connection and the irregular equisingular connections of [Connes and Marcolli (2004)].

We explain here, following [Marcolli (2008)], how regular singular Gauss–Manin connections associated to singularities of individual graph hypersurfaces X_Γ can be assembled, over different graphs, to give rise to an irregular singular connection of the kind used in [Connes and Marcolli (2004)].

In the following we let

$$\left[\frac{\omega_i}{df}\right] \qquad i = 1, \ldots, \mu \tag{8.34}$$

be a basis for the vanishing cohomology bundle, written with the same notation we used above for the Gelfand–Leray form. Then the Gauss–Manin connection on the vanishing cohomology bundle, which is defined by the integer cohomology lattice in each real cohomology fiber, acts on the basis (8.34) by

$$\nabla_s^{GM} \left[\frac{\omega_i}{df} \right]_s = \sum_j p_{ij}(s) \left[\frac{\omega_j}{df} \right]_s, \tag{8.35}$$

where the $p_{ij}(s)$ are holomorphic away from $s = 0$ and have a pole at $s = 0$. The Gauss–Manin connection is regular singular and its monodromy agrees with the monodromy of the singularity (see [Arnold, Goryunov, Lyashko, Vasilev (1998)], §2.3). Given a covariantly constant section $\delta(s)$ of the vanishing homology bundle, the function

$$I(s) = \left(\int_{\delta(s)} \frac{\omega_1}{df}, \ldots, \int_{\delta(s)} \frac{\omega_\mu}{df} \right) \tag{8.36}$$

is a solution of the regular singular Picard–Fuchs equation

$$\frac{d}{ds} I(s) = P(s) I(s), \quad \text{with} \quad P(s)_{ij} = p_{ij}(s). \tag{8.37}$$

Similarly, suppose given a holomorphic n-form ω and let ω/df be the corresponding Gelfand–Leray form, defining a section $[\omega/df]$ of the vanishing cohomology bundle. Let $\delta_1, \ldots, \delta_\mu$ be a basis of the vanishing homology, $\delta_i(s) \in H_{n-1}(F_s, \mathbb{Z})$. Then the function

$$I(s) = \left(\int_{\delta_1(s)} \frac{\omega}{df}, \ldots, \int_{\delta_\mu(s)} \frac{\omega}{df} \right) \tag{8.38}$$

satisfies a regular singular order ℓ differential equation

$$I^{(\ell)}(s) + p_1(s) I^{(\ell-1)}(s) + \cdots + p_\ell(s) I(s) = 0, \tag{8.39}$$

where the order is bounded above by the multiplicity of the critical point (see [Arnold, Gusein-Zade, Varchenko (1988)], §12.2.1). One refers to (8.39), or to the equivalent system of regular singular homogeneous first order equations

$$\frac{d}{ds} \mathcal{I}(s) = \mathcal{P}(s) \mathcal{I}(s), \tag{8.40}$$

with

$$\mathcal{I}_r(s) = s^{r-1} I^{(r-1)}(s), \tag{8.41}$$

as the Picard–Fuchs equation of ω. For the relation between Picard–Fuchs equations and mixed Hodge structures see §12 of [Arnold, Gusein-Zade, Varchenko (1988)] and [Kulikov (1998)].

We also recall some of the properties of the irregular singular, equi-singular connections of [Connes and Marcolli (2004)] which we need here.

We let $\mathfrak{g}$ denote the Lie algebra $\mathfrak{g} = \mathrm{Lie}(G)$. Let K denote the field of germs of meromorphic functions at $z = 0$. We also let B denote a fibration over an infinitesimal disk Δ^* with fiber the multiplicative group $\mathbb{G}_m$ and we denote by P the principal G-bundle $P = B \times G$. We consider $\mathrm{Lie}(G)$-valued flat connections ω that are *equisingular*. As we recalled earlier, this means that they satisfy the following conditions.

- The connections satisfy $\omega(z, \lambda u) = \lambda^Y \omega(z, u)$, for $\lambda \in \mathbb{G}_m$, with Y the grading operator.
- Solutions of $D\gamma = \omega$, have the property that their pullbacks $\sigma^*(\gamma) \in G(K)$ along any section $\sigma : \Delta \to B$ with fixed value $\sigma(0)$ have the same negative piece of the Birkhoff factorization $\sigma^*(\gamma)_-$.

The first condition and the flatness condition imply that the connection $\omega(z, u)$ can be written in the form

$$\omega(z, u) = u^Y(a(z))\, dz + u^Y(b(z))\, \frac{du}{u}, \qquad (8.42)$$

where $a(z)$ and $b(z)$ are elements of $\mathfrak{g}(K)$ satisfying the flatness condition

$$\frac{db}{dz} - Y(a) + [a, b] = 0. \qquad (8.43)$$

Recall that the Lie bracket in the Lie algebra $\mathrm{Lie}(G)$ of the Connes–Kreimer Hopf algebra is obtained by assigning (see [Connes and Kreimer (2000)], [Connes and Kreimer (2001)])

$$[\Gamma, \Gamma'] = \sum_{v \in V(\Gamma)} \Gamma \circ_v \Gamma' - \sum_{v' \in V(\Gamma')} \Gamma' \circ_{v'} \Gamma, \qquad (8.44)$$

where $\Gamma \circ_v \Gamma'$ denotes the graph obtained by inserting Γ' into Γ at the vertex $v \in V(\Gamma)$ and the sum is over all vertices where an insertion is possible. Counting all possible ways in which a given graph can be inserted into another at a specified vertex attaches multiplicities to each insertion that appear in the form of symmetry factors.

The equisingularity condition, which determines the behavior of pull-backs of solutions along sections of the fibration $\mathbb{G}_m \to B \to \Delta$, can be checked by writing the equation $Df = \omega$ in the more explicit form

$$\gamma^{-1}\frac{d\gamma}{dz} = a(z), \quad \text{and} \quad \gamma^{-1}Y(\gamma) = b(z). \qquad (8.45)$$

When one interprets elements $\gamma \in G(K)$ as algebra homomorphisms $\phi \in$ Hom$(\mathcal{H}, K)$, one can write the above equivalently in the form

$$(\phi \circ S) * \frac{d\phi}{dz} = a, \quad \text{and} \quad (\phi \circ S) * Y(\phi) = b, \qquad (8.46)$$

where S is the antipode in $\mathcal{H}$ and $*$ is the product dual to the coproduct in the Hopf algebra. This means, on generators Γ of $\mathcal{H}$,

$$\langle (\phi \circ S) \otimes \frac{d\phi}{dz}, \Delta(\Gamma) \rangle = a_\Gamma, \quad \text{and} \quad \langle (\phi \circ S) \otimes Y(\phi), \Delta(\Gamma) \rangle = b_\Gamma, \quad (8.47)$$

where

$$\Delta(\Gamma) = \Gamma \otimes 1 + 1 \otimes \Gamma + \sum_\gamma \gamma \otimes \Gamma/\gamma$$

with the sum over subdivergences, and the antipode is given inductively by

$$S(X) = -X - \sum S(X')X'', \qquad (8.48)$$

for $\Delta(X) = X \otimes 1 + 1 \otimes X + \sum X' \otimes X''$, with X' and X'' of lower degree.

We now show how to produce a flat connection of the desired form (8.42), with irregular singularities, starting from the graph hypersuraces X_Γ, a consistent choice of slicing Π_ξ, and the regular singular Picard–Fuchs equation associated to the resulting isolated singularities of $X_\Gamma \cap \Pi_\xi$.

We begin by introducing a small modification of the Hopf algebra and coproduct, which accounts for the fact of having to choose a slicing Π_ξ. This is similar to what happens when one enriches the discrete Hopf algebra by adding the data of the external momenta.

Let $\mathcal{S}_\Gamma$ denote the manifold of planes Π_ξ in $\mathbb{A}^{\#E(\Gamma)}$ with dim$\Pi_\xi \leq$ codim Sing(X_Γ). We can write $\mathcal{S}_\Gamma$ as a disjoint union

$$\mathcal{S}_\Gamma = \bigcup_{m=1}^{\text{codim Sing}(X_\Gamma)} \mathcal{S}_{\Gamma,m}, \qquad (8.49)$$

where $\mathcal{S}_{\Gamma,m}$ is the manifold of m-dimensional planes in $\mathbb{A}^{\#E(\Gamma)}$. We denote by $\mathcal{C}^\infty(\mathcal{S}_\Gamma)$ the space of test functions on $\mathcal{S}_\Gamma$ and by $\mathcal{C}_c^{-\infty}(\mathcal{S}_\Gamma)$ its dual space of distributions.

Lemma 8.4.1. *Suppose given a subgraph* $\gamma \subset \Gamma$. *Then the choice of a distribution* $\sigma \in \mathcal{C}_c^{-\infty}(\mathcal{S}_\Gamma)$ *induces distributions* $\sigma_\gamma \in \mathcal{C}_c^{-\infty}(\mathcal{S}_\gamma)$ *and* $\sigma_{\Gamma/\gamma} \in \mathcal{C}_c^{-\infty}(\mathcal{S}_{\Gamma/\gamma})$.

Proof. Given $\gamma \subset \Gamma$, neglecting external edges, we can realize the affine X_γ as a hypersurface inside a linear subspace $\mathbb{A}^{\#E(\gamma)} \subset \mathbb{A}^{\#E(\Gamma)}$ and similarly for the affine $X_{\Gamma/\gamma}$, seen as a hypersurface inside a linear subspace $\mathbb{A}^{\#E(\Gamma/\gamma)} \subset \mathbb{A}^{\#E(\Gamma)}$, where we simply identify the edges of γ or Γ/γ with a subset of the edges of the original graph Γ.

One then has a restriction map $T_\gamma : \mathcal{S}_{\Gamma,\gamma} \to \mathcal{S}_\gamma$, where $\mathcal{S}_{\Gamma,\gamma} \subset \mathcal{S}_\Gamma$ is the union of the components $\mathcal{S}_{\Gamma,m}$ of $\mathcal{S}_\Gamma$ with $m \leq \operatorname{codim} \operatorname{Sing}(X_\gamma)$,

$$\mathcal{S}_{\Gamma,\gamma} = \bigcup_{m=1}^{\operatorname{codim} \operatorname{Sing}(X_\gamma)} \mathcal{S}_{\Gamma,m}, \tag{8.50}$$

which is given by

$$T_\gamma(\Pi_\xi) = \Pi_\xi \cap \mathbb{A}^{\#E(\gamma)}. \tag{8.51}$$

This induces a map $T_\gamma : \mathcal{C}^\infty(\mathcal{S}_\gamma) \to \mathcal{C}^\infty(\mathcal{S}_\Gamma)$ given by

$$T_\gamma(f)(\Pi_\xi) = \begin{cases} f(T_\gamma(\Pi_\xi)) & \Pi_\xi \in \mathcal{S}_{\Gamma,\gamma} \\ 0 & \text{otherwise.} \end{cases} \tag{8.52}$$

In turn, this defines a map $T_\gamma : \mathcal{C}_c^{-\infty}(\mathcal{S}_\Gamma) \to \mathcal{C}_c^{-\infty}(\mathcal{S}_\gamma)$, at the level of distributions, by

$$T_\gamma(\sigma)(f) = \sigma(T_\gamma(f)). \tag{8.53}$$

The argument for Γ/γ is analogous. One sets $\sigma_\gamma = T_\gamma(\sigma_\Gamma)$ and $\sigma_{\Gamma/\gamma} = T_{\Gamma/\gamma}(\sigma_\Gamma)$. $\qquad \square$

We then enrich the original Hopf algebra $\mathcal{H}$ by adding the datum of the slicing Π_ξ. We consider the commutative algebra

$$\tilde{\mathcal{H}} = \operatorname{Sym}(\mathcal{C}_c^{-\infty}(\mathcal{S})), \tag{8.54}$$

where $\mathcal{S} = \cup_\Gamma \mathcal{S}_\Gamma$, endowed with the coproduct

$$\Delta(\Gamma, \sigma) = (\Gamma, \sigma) \otimes 1 + 1 \otimes (\Gamma, \sigma) + \sum_\gamma (\gamma, \sigma_\gamma) \otimes (\Gamma/\gamma, \sigma_{\Gamma/\gamma}). \tag{8.55}$$

The following result is then proved by the same argument used in [Connes and Marcolli (2008)], Theorem 1.27.

Lemma 8.4.2. *The coproduct* (8.55) *is coassociative and* $\tilde{\mathcal{H}}$ *is a Hopf algebra.*

We then proceed as follows. We pass to the projective instead of the affine formulation and we fix a small neighborhood of an isolated singular point of $X_\Gamma \cap \Pi_\xi$, for Π_ξ a linear space of dimension at most equal to the codimension of $\mathrm{Sing}(X_\Gamma)$. Suppose given a holomorphic k-form α_ξ on Π_ξ. Then there exists an associated regular singular Picard–Fuchs equation

$$J_{\Gamma,\xi}^{(\ell)}(s) + p_1(s)J_{\Gamma,\xi}^{(\ell-1)}(s) + \cdots + p_\ell(s)J_{\Gamma,\xi}(s) = 0, \qquad (8.56)$$

with the property that any solution $J_{\Gamma,\xi}(s)$ is a linear combination of the functions

$$J_{\Gamma,\xi,i}(s) = \int_{\delta_i(s)} \frac{\alpha_\xi}{df}, \qquad (8.57)$$

where $\delta_1, \ldots, \delta_\mu$ is a basis of locally constant sections of the homological Milnor fibration, $\delta_i(s) \in H_{k-1}(F_s, \mathbb{Z})$, and α_ξ/df is the Gelfand–Leray form associated to the holomorphic k-form α_ξ.

This depends on the choice of a singular point and can be localized in a small neighborhood of the singular point in $X_\Gamma \cap \Pi_\xi$. In fact, introducing a cutoff χ_ξ that is supported near the singularities of $X_\Gamma \cap \Pi_\xi$ amounts to adding the expressions (8.57) for the different singular points. Thus, to simplify notation, we can just assume that we have a single expression $J_{\Gamma,\xi}(s)$ at a unique isolated critical point.

We then have the following result, which constructs irregular singular connections from solutions of the regular singular Picard–Fuchs equation.

Theorem 8.4.3. *Any solution $J_{\Gamma,\xi}$ of the regular singular Picard–Fuchs equation (8.56) determines a flat $\mathfrak{g}(K)$-valued connection $\omega(z, u)$ of the form (8.42). Moreover, if the k-form α_ξ is given by $P_\Gamma^\ell \Omega_\xi$ as above, then the connection is equisingular.*

Proof. We consider the Mellin transform

$$\mathcal{F}_{\Gamma,\xi}(z) = \int_0^\infty s^z \, J_{\Gamma,\xi}(s) \, ds. \qquad (8.58)$$

As in §7 of [Arnold, Gusein-Zade, Varchenko (1988)], the function $\mathcal{F}_{\Gamma,\xi}(z)$ admits an analytic continuation to a meromorphic function with poles at points $z = -(\lambda+1)$ with $\lambda \in \Xi_{\Gamma,\xi}$ a discrete set in $\mathbb{R}$ of points related to the multiplicities of an embedded resolution of the singular point of $X_\Gamma \cap \Pi_\xi$. We look at the function $\mathcal{F}_{\Gamma,\xi}(z)$ in a small neighborhood of a chosen point $z = -D$. It has an expansion as a Laurent series, with a pole at $z = -D$ if $-D \in \Xi_{\Gamma,\xi}$.

After a change of variables on the complex coordinate z, so that we have $z \in \Delta^*$ a small neighborhood of $z = 0$, we define

$$\phi_\mu(\Gamma, \sigma)(z) := \mu^{-z\, b_1(\Gamma)} \sigma \left(\mathcal{F}_{\Gamma,\xi}(-\frac{D+z}{2}) \right), \qquad (8.59)$$

where we consider $\mathcal{F}_{\Gamma,\xi}$ as a function of ξ to which we apply the distribution σ. More precisely, after identifying $\mathcal{F}_{\Gamma,\xi}$ with its Laurent series expansion, we apply σ to the coefficients seen as functions of ξ. This defines an algebra homomorphism $\phi_\mu \in \mathrm{Hom}(\tilde{\mathcal{H}}, K)$, by assigning the values (8.59) on generators. Here μ is the mass scale. The homomorphism ϕ defined by (8.59) can be equivalently described as a family of $\tilde{G}(\mathbb{C})$-valued loops $\gamma_\mu : \Delta^* \to \tilde{G}(\mathbb{C})$, depending on the mass scale μ. Here $\tilde{G}$ denotes the affine group scheme dual to the commutative Hopf algebra $\tilde{\mathcal{H}}$. The dependence on μ of (8.59) implies that γ_μ satisfies the scaling property

$$\gamma_{e^t \mu}(z) = \theta_{tz}(\gamma_\mu(z)), \qquad (8.60)$$

where θ_t is the one-parameter family of automorphisms of $\tilde{\mathcal{H}}$ generated by the grading, $\frac{d}{dt}\theta_t|_{t=0} = Y$. Then one sets

$$a_\mu(z) := (\phi_\mu \circ S) * \frac{d}{dz}\phi_\mu, \quad \text{and} \quad b_\mu(z) := (\phi_\mu \circ S) * Y(\phi_\mu), \qquad (8.61)$$

where S and $*$ are the antipode of $\tilde{\mathcal{H}}$ and the product dual to the coproduct Δ of (8.55). These define elements a_μ, b_μ $\Omega^1(\mathfrak{g}(K))$, which one can use to define a connection $\omega(z, u)$ of the form (8.42). More precisely, for $\mu = e^t$, one has

$$\gamma_\mu^{-1} \frac{d}{dz} \gamma_\mu = \theta_t(\gamma^{-1}\frac{d}{dz}\gamma) = u^Y(a(z)),$$

where we set $u^Y = e^{tY}$ and then extend the resulting expression to $u \in \mathbb{G}_m(\mathbb{C}) = \mathbb{C}^*$. Similarly, we get $\gamma_\mu^{-1} Y(\gamma_\mu) = u^Y(b(z))$. Thus, the connection $\omega(z, u)$ defined in this way satisfies by construction the first condition of the equisingularity property, namely $\omega(z, \lambda u) = \lambda^Y \omega(z, u)$, for all $\lambda \in \mathbb{G}_m$. One can see that the connection is flat since we have

$$\frac{d}{dz}b_\mu(z) - Y(a_\mu(z)) = \frac{d\gamma_\mu^{-1}(z)}{dz}Y(\gamma_\mu(z)) + \gamma_\mu^{-1}(z)\frac{d}{dz}(Y(\gamma_\mu(z)))$$

$$-Y(\gamma_\mu^{-1}(z))\frac{d}{dz}\gamma_\mu(z) - \gamma_\mu^{-1}(z)\frac{d}{dz}(Y(\gamma_\mu(z)))$$

$$= -\gamma_\mu^{-1}(z)\frac{d}{dz}(\gamma_\mu(z))\gamma_\mu^{-1}(z)Y(\gamma_\mu(z)) - \gamma_\mu^{-1}(z)Y(\gamma_\mu(z))\gamma_\mu^{-1}(z)$$

$$= -[a(z), b(z)].$$

The second condition of equisingularity is the property that, in the Birkhoff factorization

$$\gamma_\mu(z) = \gamma_{\mu,-}(z)^{-1}\gamma_{\mu,+}(z),$$

the negative part satisfies

$$\frac{d}{d\mu}\gamma_{\mu,-}(z) = 0.$$

By dimensional analysis on the counterterms, in the case of dimensional regularization and minimal subtraction, it is possible to show (see [Collins (1986)] §5.8.1) that the counterterms obtained by the BPHZ procedure applied to the Feynman integral $U_\mu(\Gamma)(z)$ do not depend on the mass parameter μ. This means, as shown in [Connes and Kreimer (2000)] (see also Proposition 1.44 of [Connes and Marcolli (2008)]), that the Feynman integrals $U_\mu(\Gamma)(z)$ define a $G(\mathbb{C})$-valued loop $\gamma_\mu(z)$ with the property that $\partial_\mu\gamma_{\mu,-}(z) = 0$. The integrals (8.58) considered here correspond to slices along a linear space Π_ξ of the Feynman integrals, localized by a cutoff χ_ξ near the singular points. The explicit dependence on μ in the Feynman integrals is as in (8.59), which is unchanged with respect to that of the original dimensionally regularized Feynman integrals. Thus, the same argument as in [Collins (1986)] §5.8.1 and Proposition 1.44 of [Connes and Marcolli (2008)] applies to this case to show that $\partial_\mu\gamma_{\mu,-}(z) = 0$. $\qquad\square$

Chapter 9

Beyond scalar theories

In this final chapter we give a few examples of how the theme of Feynman integrals and motives, based on the parametric representation of Feynman integrals and the associated algebraic varieties, can be extended beyond the original setting of scalar quantum field theories. We concentrate here only on two simple kinds of extensions, where one can already see some new feature appearing. The first extension we discuss is to theories with fermionic fields, based on the results of [Marcolli and Rej (2008)]. In this case one can again express a parametric form of the Feynman integral in terms of periods, this time on a supermanifold, with the determinant defining the graph polynomial replaced by a Berezinian, which is a natural analog in the supergeometry context. The second case we discuss is that of scalar quantum field theories, and in particular the ϕ^4 theory, over a spacetime that is deformed to a noncommutative manifold with a Moyal product. In this case it is also known that a parametric formulation of Feynman integrals exists, and we report some observations of [Aluffi and Marcolli (2008a)] on computations of characteristic classes for the resulting graph hypersurfaces.

9.1 Supermanifolds

We describe here an extension of the parametric Feynman integral and its motivic interpretation to the case of theories with both fermionic and bosonic variables, where we can reformulate the parametric integral as a period of a supermanifold, defined in terms of Berezinians instead of determinants and fermionic integration. The results reported here are from [Marcolli and Rej (2008)].

We report a few basic facts about the geometry of supermanifolds, following [Manin (1988)].

By a complex supermanifold one understands a datum $\mathcal{X} = (X, \mathcal{A})$ with the following properties: $\mathcal{A}$ is a sheaf of supercommutative rings on X; $(X, \mathcal{O}_X)$ is a complex manifold, where $\mathcal{O}_X = \mathcal{A}/\mathcal{N}$, with $\mathcal{N}$ the ideal of nilpotents in $\mathcal{A}$; the quotient $\mathcal{E} = \mathcal{N}/\mathcal{N}^2$ is locally free over $\mathcal{O}_X$ and $\mathcal{A}$ is locally isomorphic to the exterior algebra $\Lambda_{\mathcal{O}_X}^\bullet(\mathcal{E})$, where the grading is the $\mathbb{Z}/2\mathbb{Z}$-grading by odd/even degrees. The supermanifold is split if the isomorphism $\mathcal{A} \cong \Lambda_{\mathcal{O}_X}^\bullet(\mathcal{E})$ is global. Thus, in local coordinates, a supermanifold is parameterized by bosonic and fermionic variables $(x_0, \ldots, x_n; \theta_1, \ldots, \theta_m)$, where the x_i are ordinary, commuting, scalar variables, while the θ_j are anticommuting Grassmann variables. The bosonic variables x_i generate the order zero, or even degree, part and the fermionic variables θ_j the order one, or odd degree part with respect to the $\mathbb{Z}/2\mathbb{Z}$-grading.

A simple example is given by the projective superspaces. The complex projective superspace $\mathbb{P}^{n|m}$ is the supermanifold $(X, \mathcal{A})$ with $X = \mathbb{P}^n$ the usual complex projective space and

$$\mathcal{A} = \Lambda^\bullet(\mathbb{C}^m \otimes_{\mathbb{C}} \mathcal{O}(-1)),$$

with the exterior powers $\Lambda^\bullet$ graded by odd/even degree. It is a split supermanifold.

A *morphism* $F : \mathcal{X}_1 \to \mathcal{X}_2$ of supermanifolds $\mathcal{X}_i = (X_i, \mathcal{A}_i)$, $i = 1, 2$, consists of a pair $F = (f, f^\#)$ of a morphism of the underlying complex manifolds $f : X_1 \to X_2$ together with a morphism $f^\# : \mathcal{A}_2 \to f_* \mathcal{A}_1$ of sheaves of supercommutative rings with the property that at each point $x \in X_1$ the induced morphism $f_x^\# : (\mathcal{A}_2)_{f(x)} \to (\mathcal{A}_1)_x$ satisfies $f_x^\#(\mathfrak{m}_{f(x)}) \subset \mathfrak{m}_x$, on the maximal ideals of germs of sections vanishing at the point (*cf.* [Manin (1988)], §4.1).

In particular, an *embedding* of complex supermanifolds is a morphism $F = (f, f^\#)$ as above, with the property that $f : X_1 \hookrightarrow X_2$ is an embedding and $f^\# : \mathcal{A}_2 \to f_* \mathcal{A}_1$ is surjective. As in ordinary geometry, we define the ideal sheaf of $\mathcal{X}_1$ to be the kernel

$$\mathcal{I}_{\mathcal{X}_1} := \mathrm{Ker}(f^\# : \mathcal{A}_2 \to f_* \mathcal{A}_1). \tag{9.1}$$

An equivalent characterization of an embedding of supermanifold is given as follows. If we denote by E_i, for $i = 1, 2$ the holomorphic vector bundles on X_i such that $\mathcal{O}(E_i) = \mathcal{E}_i = \mathcal{N}_i/\mathcal{N}_i^2$, with the notation as above, then an embedding $F : \mathcal{X}_1 \hookrightarrow \mathcal{X}_2$ is an embedding $f : X_1 \hookrightarrow X_2$ such that the induced morphism of vector bundles $f^* : E_2 \to E_1$ is surjective. Thus, we say that $\mathcal{Y} = (Y, \mathcal{B})$ is a closed sub-supermanifold of $\mathcal{X} = (X, \mathcal{A})$ when there exists a closed embedding $Y \subset X$ and the pullback of $E_{\mathcal{A}}$ under this embedding surjects to $E_{\mathcal{B}}$.

An open submanifold $\mathcal{U} = (U, \mathcal{B}) \hookrightarrow \mathcal{X} = (X, \mathcal{A})$ is given by an open embedding $U \hookrightarrow X$ of the underlying complex manifolds and an isomorphism of sheaves $\mathcal{A}|_U \cong \mathcal{B}$. When $\mathcal{Y} \subset \mathcal{X}$ is a closed embedding and $U = X \smallsetminus Y$, the ideal sheaf of $\mathcal{Y}$ satisfies $\mathcal{I}_{\mathcal{Y}}|_U = \mathcal{A}|_U$.

A subvariety in superprojective space is a supermanifold

$$\mathcal{X} = (X \subset \mathbb{P}^n, (\Lambda^\bullet(\mathbb{C}^m \otimes_{\mathbb{C}} \mathcal{O}(-1))/\mathcal{I})|_X), \tag{9.2}$$

where $\mathcal{I} = \mathcal{I}_{\mathcal{X}}$ is an ideal generated by finitely many homogeneous polynomials of given $\mathbb{Z}/2$-parity. In other words, if we denote by $(x_0, \ldots, x_n, \theta_1, \ldots, \theta_m)$ the bosonic and fermionic coordinates of $\mathbb{P}^{n|m}$, then a projective subvariety can be obtained by assigning a number of equations of the form

$$\Psi^{\mathrm{ev/odd}}(x_0, \ldots, x_n, \theta_1, \ldots, \theta_m) = \sum_{i_1 < \cdots < i_k} P_I(x_0, \ldots, x_n)\theta_{i_1} \cdots \theta_{i_k} = 0,$$
$$\tag{9.3}$$

where $I = (i_1, \ldots, i_k)$, and the $P_I(x_0, \ldots, x_n)$ are homogeneous polynomials in the bosonic variables.

In [Marcolli and Rej (2008)] an analog of the Grothendieck ring of varieties is considered for supermanifolds, defined as follows.

Definition 9.1.1. Let $\mathcal{SV}_{\mathbb{C}}$ be the category of complex supermanifolds with morphisms defined as above. Let $K_0(\mathcal{SV}_{\mathbb{C}})$ denote the free abelian group generated by the isomorphism classes of objects $\mathcal{X} \in \mathcal{SV}_{\mathbb{C}}$ subject to the following relations. Let $F : \mathcal{Y} \hookrightarrow \mathcal{X}$ be a closed embedding of supermanifolds. Then

$$[\mathcal{X}] = [\mathcal{Y}] + [\mathcal{X} \smallsetminus \mathcal{Y}], \tag{9.4}$$

where $\mathcal{X} \smallsetminus \mathcal{Y}$ is the supermanifold

$$\mathcal{X} \smallsetminus \mathcal{Y} = (X \smallsetminus Y, \mathcal{A}_X|_{X \smallsetminus Y}). \tag{9.5}$$

Here the notation $\mathcal{A}_X$ means the following, see [Kashiwara and Schapira (1990)] §II.2.3. Given a locally closed subset $Y \subset X$ and a sheaf $\mathcal{A}$ on X, there exists a unique sheaf $\mathcal{A}_Y$ with the property that

$$\mathcal{A}_Y|_Y = \mathcal{A}|_Y \quad \text{and} \quad \mathcal{A}_Y|_{X \smallsetminus Y} = 0. \tag{9.6}$$

In the case where Y is closed, this satisfies $\mathcal{A}_Y = i_*(\mathcal{A}|_Y)$ where $i : Y \hookrightarrow X$ is the inclusion. When Y is open it satisfies $\mathcal{A}_Y = \mathrm{Ker}(\mathcal{A} \to j_*(\mathcal{A}|_{X \smallsetminus Y}))$, where in this case $j : X \smallsetminus Y \hookrightarrow X$ is the closed embedding of the complement.

In particular, in the case where $\mathcal{A} = \mathcal{O}_X$ is the structure sheaf of X, the relation (9.4) reduces to the usual relation

$$[X] = [Y] + [X \smallsetminus Y]. \tag{9.7}$$

in the Grothendieck group of ordinary varieties, for a closed embedding $Y \subset X$.

Proposition 9.1.2. *The Grothendieck ring $K_0(\mathcal{SV}_{\mathbb{C}})$ of supervarieties is a polynomial ring over the Grothendieck ring of ordinary varieties of the form*

$$K_0(\mathcal{SV}_{\mathbb{C}}) = K_0(\mathcal{V}_{\mathbb{C}})[T], \tag{9.8}$$

where $T = [\mathbb{A}^{0|1}]$ is the class of the affine superspace of dimension $(0, 1)$.

Proof. We show that all supermanifolds decompose in $K_0(\mathcal{SV}_{\mathbb{C}})$ as a finite combination of split supermanifolds, and in fact of supermanifolds where the vector bundle E with $\mathcal{O}(E) = \mathcal{E} = \mathcal{N}/\mathcal{N}^2$ is trivial. This is a consequence of the dévissage of coherent sheaves, see [Shafarevich (1996)], §3.3, Propositions 1–3. Namely, for any coherent sheaf $\mathcal{A}$ over a Noetherian reduced irreducible scheme there exists a dense open set U such that $\mathcal{A}|_U$ is free. The relation (9.4) then ensures that, given a pair $\mathcal{X} = (X, \mathcal{A})$ and the sequence of sheaves

$$0 \to i_!(\mathcal{A}|_U) \to \mathcal{A} \to j_*(\mathcal{A}|_Y) \to 0,$$

associated to the open embedding $U \subset X$ with complement $Y = X \smallsetminus U$, the class $[X, \mathcal{A}]$ satisfies

$$[X, \mathcal{A}] = [U, \mathcal{A}_U|_U] + [Y, \mathcal{A}_Y|_Y].$$

The sheaf $\mathcal{A}_Y$ on X, which has support Y, has a chain of subsheaves $\mathcal{A}_Y \supset \mathcal{A}_1 \supset \cdots \supset \mathcal{A}_k = 0$ such that each quotient $\mathcal{A}_i/\mathcal{A}_{i+1}$ is coherent on Y, [Shafarevich (1996)], §3.3, Proposition 3. Thus, one can find a stratification where on each open stratum the supermanifold is split and with trivial vector bundle. The supermanifold $\mathcal{X} = (X, \mathcal{A})$ decomposes as a sum of the corresponding classes in the Grothendieck group. It is then clear that the product makes $K_0(\mathcal{SV}_{\mathbb{C}})$ into a ring with

$$[\mathcal{X}][\mathcal{Y}] = [\mathcal{X} \times \mathcal{Y}],$$

and the stated result then follows. The fact that it is indeed a polynomial ring in the class $[\mathbb{A}^{0|1}]$ without further relation is a consequence of the fact that $\mathbb{A}^{0|1}$ is parameterized by a single Grassmann variable and that the $\mathbb{A}^{0|n}$ are non-isomorphic supermanifolds. $\quad\square$

The fact that the vector bundle that constitutes the fermionic part of a supermanifold is trivial when seen in the Grothendieck group is the analog for supermanifolds of the fact that any locally trivial fibration is equivalent to a product in the Grothendieck group of ordinary varieties.

Notice that, in the supermanifold case, there are now two different types of Lefschetz motives, namely the bosonic one $\mathbb{L}_b = [\mathbb{A}^{1|0}]$ and the fermionic one $\mathbb{L}_f = [\mathbb{A}^{0|1}]$. The formal inverses of $\mathbb{L}_f$ and $\mathbb{L}_b$ correspond to two types of Tate objects for motives of supermanifolds, respectively fermionic and bosonic Tate motives. The contribution of the fermionic part of a supermanifold to the class in the Grothendieck ring is always of (fermonic) Tate type, while only the bosonic part can provide non-Tate contributions.

The analog of the determinant in supergeometry is given by the Berezinian. This is defined in the following way. Suppose given a matrix $\mathcal{M}$ of the form

$$\mathcal{M} = \begin{pmatrix} M_{11} & M_{12} \\ M_{21} & M_{22} \end{pmatrix},$$

where the M_{11} and M_{22} are square matrices with entries of order zero and the M_{12} and M_{21} have elements of order one. Then the Berezinian of $\mathcal{M}$ is the expression

$$\mathrm{Ber}(\mathcal{M}) := \frac{\det(M_{11} - M_{12}M_{22}^{-1}M_{21})}{\det(M_{22})}. \tag{9.9}$$

It satisfies $\mathrm{Ber}(\mathcal{M}\mathcal{N}) = \mathrm{Ber}(\mathcal{M})\mathrm{Ber}(\mathcal{N})$.

Grassmann variables integration is defined in terms of the Berezinian integral, which is determined by the property that, for a form $\omega = \sum_I f_I(x_0, \ldots, x_n)\theta^I$ in the local coordinates $(x_0, \ldots, x_n; \theta_1, \ldots, \theta_m)$, the integral gives (see Chapter 4, §6 of [Manin (1988)])

$$\int \omega = \int f_{1,\ldots,m}(x_0, \ldots, x_n)\, dx_1 \cdots dx_n, \tag{9.10}$$

or equivalently by the property that

$$\int (h_0(x) + h_1(x)\theta)\, d\theta = h_1(x), \tag{9.11}$$

for x an ordinary and θ a Grassmann variable. It is shown in [Bernšteĭn and Leĭtes (1977)] that Grassmann integration satisfies a change of variable formula where the Jacobian of the coordinate change is given by the Berezinian Ber(J) with J the matrix $J = \frac{\partial X_\alpha}{\partial Y_\beta}$ with $X_\alpha = (x_i, \xi_r)$ and $Y_\beta = (y_j, \eta_s)$.

There is also a well developed theory of divisors on supermanifolds, originating from [Rosly, Schwarz, Voronov (1988)]. A Cartier divisor on a supermanifold $\mathcal{X}$ of dimension $(n|m)$ is defined by a collection of *even* meromorphic functions ϕ_i defined on an open covering $\mathcal{U}_i \hookrightarrow \mathcal{X}$, with $\phi_i \phi_j^{-1}$ a holomorphic function on $\mathcal{U}_i \cap \mathcal{U}_j$ nowhere vanishing on the underlying $U_i \cap U_j$. Classes of divisors correspond to equivalence classes of line bundles and can be described in terms of integer linear combinations of $(n-1|m)$-dimensional subvarieties $\mathcal{Y} \subset \mathcal{X}$.

9.2   Parametric Feynman integrals and supermanifolds

Consider now the case of Feynman propagators and Feynman diagrams that come from theories with both bosonic and fermionic fields. It is well known that for such theories the form of the possible fermionic terms in the Lagrangian is severely constrained: for instance, in dimension $D = 4$ renormalizable interaction terms can only involve at most two fermion and one boson field, see [Ramond (1990)], §5.3.

The perturbative expansion for such theories correspondingly involve graphs Γ with two different types of edges: fermionic and bosonic edges. The Feynman rules assign to each bosonic edge a propagator of the usual form we have encountered so far, and to fermionic edges a propagator

$$i\frac{\not{p} + m}{p^2 - m^2} = \frac{i}{\not{p} - m}. \tag{9.12}$$

More generally, in cases with tensor indices, we would have $\not{p} = p^\mu \gamma_\mu$, involving γ-matrices expressing fermions in spinorial form.

In the following we use the notation

$$q(p) = p^2 - m^2, \qquad \not{q}(p) = i(\not{p} + m) \tag{9.13}$$

for the quadratic and linear forms that appear in the bosonic and fermionic propagators. In the following, again just to simplify notation, we also drop the mass terms in the propagator, *i.e.* we set $m = 0$, and ignore, for simplicity, the resulting infrared divergence problem.

Thus, the terms of the form $(q_1 \cdots q_n)^{-1}$, which we encountered in the purely bosonic case, are now replaced by terms of the form

$$\frac{\not q_1 \cdots \not q_f}{q_1 \cdots q_n}, \tag{9.14}$$

where $n = \#E(\Gamma)$ is the total number of edges in the graph and $f = \#E_f(\Gamma)$ is the number of fermionic edges.

We have an analog of the Feynman trick we saw in the first chapter, where we now introduce an extra set of Grassmann variables associated to the fermionic edges. The derivation we present suffers from a kind of "fermion doubling problem", as each fermionic edge contributes an ordinary integration variable, which essentially account for the denominator q_i in (9.12) and (9.14), as well as a *pair* of Grassmann variables accounting for the numerator $\not q_i$ in (9.12) and (9.14). Other formulations of parametric Feynman integrals for theories with fermionic variables are possible, for example as done in [Medvedev (1978)].

Let $\mathcal{Q}_f$ denote the $2f \times 2f$ antisymmetric matrix

$$\mathcal{Q}_f = \begin{pmatrix} 0 & \not q_1 & 0 & 0 & \cdots & 0 & 0 \\ -\not q_1 & 0 & 0 & 0 & \cdots & 0 & 0 \\ 0 & 0 & 0 & \not q_2 & \cdots & 0 & 0 \\ 0 & 0 & -\not q_2 & 0 & \cdots & 0 & 0 \\ \vdots & & & & \cdots & & \vdots \\ 0 & 0 & 0 & 0 & \cdots & 0 & \not q_f \\ 0 & 0 & 0 & 0 & \cdots & -\not q_f & 0 \end{pmatrix}. \tag{9.15}$$

Lemma 9.2.1. *Let $\sigma_{n|2f}$ denote the superspace $\sigma_n \times \mathbb{A}^{0|2f}$. Then the following identity holds:*

$$\frac{\not q_1 \cdots \not q_f}{q_1 \cdots q_n} = K_{n,f} \int_{\sigma_{n|2f}} \frac{dt_1 \cdots dt_{n-1} d\theta_1 \cdots d\theta_{2f}}{(t_1 q_1 + \cdots t_n q_n + \frac{1}{2}{}^t\theta \mathcal{Q}_f \theta)^{n-f}}, \tag{9.16}$$

with

$$K_{n,f} = \frac{2^f (n-1)!}{\prod_{k=1}^{f}(-n+f-k+1)}.$$

Proof. We first show that the following identity holds for integration in the Grassmann variables $\theta = (\theta_1, \ldots, \theta_{2f})$:

$$\int \frac{d\theta_1 \cdots d\theta_{2f}}{(1 + \frac{1}{2}{}^t\theta \mathcal{Q}_f \theta)^{\alpha}} = \frac{f!}{2^f} \binom{-\alpha}{f} \not q_1 \cdots \not q_f. \tag{9.17}$$

In fact, we expand using the Taylor series

$$(1+x)^\beta = \sum_{k=0}^\infty \binom{\beta}{k} x^k$$

and the identity

$$\frac{1}{2}\theta^t \mathcal{Q}_f \theta = \sum_{i=1}^f q_i \theta_{2i-1}\theta_{2i},$$

together with the fact that the degree zero variables $x_i = \theta_{2i-1}\theta_{2i}$ commute, to obtain

$$(1 + \frac{1}{2}\theta^t \mathcal{Q}_f \theta)^{-\alpha} = \sum_{k=0}^\infty \binom{-\alpha}{k} (\sum_{i=1}^f q_i \theta_{2i-1}\theta_{2i})^k.$$

The rules of Grassmann integration then imply that only the coefficient of $\theta_1 \cdots \theta_{2f}$ remains as a result of the integration. This gives (9.17).

For simplicity of notation, we then write $T = t_1 q_1 + \cdots t_n q_n$, so that we have

$$\int_{\sigma_{n|2f}} \frac{1}{(t_1 q_1 + \cdots t_n q_n + \frac{1}{2}\theta^t \mathcal{Q}_f \theta)^{n-f}} dt_1 \cdots dt_{n-1}\, d\theta_1 \cdots d\theta_{2f}$$

$$= \frac{f!}{2^f} \binom{-n+f}{f} q_1 \cdots q_f \int_{\sigma_n} T^{-n+f} T^{-f}\, dt_1 \cdots dt_{n-1}$$

$$= \frac{f!}{2^f} \binom{-n+f}{f} q_1 \cdots q_f \int_{\sigma_n} \frac{dt_1 \cdots dt_{n-1}}{(t_1 q_1 + \cdots + t_n q_n)^n}$$

$$= \frac{f!}{2^f (n-1)!} \binom{-n+f}{f} \frac{q_1 \cdots q_f}{q_1 \cdots q_n}.$$

$\qquad\qquad\qquad\qquad\qquad\qquad\qquad\qquad\qquad\qquad\qquad\qquad\qquad\square$

Consider now the case of graphs that have both bosonic and fermionic legs. We mimic the procedure described earlier in this chapter to obtain a parametric form for the Feynman integral. We will deal here with a particular class of graph, which can be considered an equivalent, in the case with fermionic variables, of the log divergent graphs, for which the parametric form of the Feynman integral is simplified by the lack of the second graph polynomial (see §3.1). As we are going to see below, here the condition that characterizes the class of graphs with this property will no longer depend only on the number of internal edges and loops but also on a particular choice of a basis of loops.

Theorem 9.2.2. *Suppose given a graph* Γ *with* n *edges, of which* f *are fermionic and* $b = n - f$ *bosonic. Assume that there exists a choice of a basis for* $H_1(\Gamma)$ *satisfying the condition*

$$n - \frac{f}{2} + \frac{D}{2}(\ell_f - \ell_b) = 0. \tag{9.18}$$

Then the following identity holds:

$$\int \frac{dq_1 \cdots dq_f}{q_1 \cdots q_n} d^D s_1 \cdots d^D s_{\ell_b} \, d^D \sigma_1 \cdots d^D \sigma_{\ell_f} = \int_{\sigma_n} \frac{\Lambda(t)}{\mathrm{Ber}(\mathcal{M}(t))^{D/2}} dt_1 \cdots dt_n. \tag{9.19}$$

Proof. We divide the edge indices $i = 1, \ldots, n$ of the Feynman graph into two sets $i_b = 1, \ldots, n_b$ and $i_f = 1, \ldots, n_f$, with $n = n_b + n_f$, respectively labeling the bosonic and fermionic legs. Consequently, given a choice of a basis for the first homology of the graph, indexed as above by $r = 1, \ldots, \ell$, we replace the circuit matrix η_{ir} of (3.1) with a matrix of the form

$$\begin{pmatrix} \eta_{i_f r_f} & \eta_{i_f r_b} \\ \eta_{i_b r_f} & \eta_{i_b r_b} \end{pmatrix}. \tag{9.20}$$

Here the loop indices $r = 1, \ldots, \ell$ are at first divided into three sets $\{1, \ldots, \ell_{ff}\}$, labelling the loops consisiting of only fermionic edges, $\{1, \ldots, \ell_{bb}\}$ labelling the loops consisting of only bosonic edges, and the remaning variables $\{1, \ldots, \ell_{bf} = \ell - (\ell_{ff} + \ell_{bb})\}$ for the loops that contain both fermionic and bosonic edges. We then introduce two sets of momentum variables: ordinary variables $\displaystyle{\not}s_{r_b} \in \mathbb{A}^{D|0}$, with $r_b = 1, \ldots, \ell_b = \ell_{bb} + \ell_{bf}$, and Grassmann variables $\sigma_{r_f} \in \mathbb{A}^{0|D}$ with $r_f = 1, \ldots, \ell_f = \ell_{ff} + \ell_{bf}$. That is, we assign to each purely fermionic loop a Grassmann momentum variable, to each purely bosonic loop an ordinary momentum variable, and to the loops containing both fermionic and bosonic legs a pair $(\not{s}_r, \sigma_r)$ of an ordinary and a Grassman variable. In (9.20) above we write r_f and r_b, respectively, for the indexing sets of these Grassmann and ordinary variables.

We then consider a change of variables

$$\not{p}_{i_b} = \not{u}_{i_b} + \sum_{r_f} \eta_{i_b r_f} \sigma_{r_f} + \sum_{r_b} \eta_{i_b r_b} \not{s}_{r_b}, \quad \not{p}_{i_f} = \not{u}_{i_f} + \sum_{r_f} \eta_{i_f r_f} \sigma_{r_f} + \sum_{r_b} \eta_{i_f r_b} \not{s}_{r_b}. \tag{9.21}$$

analogous to the one used before, where now, for reasons of homogeneity, we need to assume that the $\eta_{i r_f}$ are of degree one and the $\eta_{i r_b}$ are of degree zero, since the $\not{p}_i$ are even (ordinary) variables.

We apply the change of variables (9.21) to the expression

$$\sum_i t_i p_i^2 + \sum_{i_f} \theta_{2i_f - 1} \theta_{2i_f} \not{p}_{i_f}. \tag{9.22}$$

We assume again, as in the purely bosonic case (*cf.* (18.35) of [Bjorken and Drell (1965)]), the relations

$$\sum_i t_i \psi_i \eta_{ir} = 0$$

for each loop variable $r = r_b$ and $r = r_f$.

We can then rewrite (9.22) in the form

$$\sum_i t_i u_i^2 + \sum_{i_f} \theta_{2i_f-1} \theta_{2i_f} \psi_{i_f}$$

$$+ \sum_{r_b, r_b'} (\sum_i t_i \eta_{ir_b} \eta_{ir_b'}) \not{s}_{r_b} \not{s}_{r_b'} - \sum_{r_f r_f'} (\sum_i t_i \eta_{ir_f} \eta_{ir_f'}) \sigma_{r_f} \sigma_{r_f'}$$

$$+ \sum_{r_b r_f} \left((\sum_i t_i \eta_{ir_b} \eta_{ir_f}) \not{s}_{r_b} \sigma_{r_f} - \sigma_{r_f}^\tau \not{s}_{r_b}^\tau (\sum_i t_i \eta_{ir_f} \eta_{ir_b}) \right)$$

$$+ \sum_{r_b} (\sum_{i_f} \theta_{2i_f-1} \theta_{2i_f} \eta_{i_f r_b}) \not{s}_{r_b} + \sum_{r_f} (\sum_{i_f} \theta_{2i_f-1} \theta_{2i_f} \eta_{i_f r_f}) \sigma_{r_f}.$$

Notice the minus sign in front of the quadratic term in the σ_{r_f}, since for order-one variables $\sigma_{r_f} \eta_{ir_f'} = -\eta_{ir_f'} \sigma_{r_f}$. We write the above in the simpler notation

$$T + \not{s}^\tau M_b(t) \not{s} - \sigma^\tau M_f(t) \sigma + \sigma^\tau M_{fb}(t) \not{s} - \not{s}^\tau M_{bf}(t) \sigma + N_b(\theta) \not{s} + N_f(\theta) \sigma, \tag{9.23}$$

where τ denotes transposition, $\not{s} = (\not{s}_{r_b})$, $\sigma = (\sigma_{r_f})$, and

$$\begin{aligned}
T &= \sum_i t_i u_i^2 + \sum_{i_f} \theta_{2i_f-1} \theta_{2i_f} \psi_{i_f}, \\
M_b(t) &= \sum_i t_i \eta_{ir_b} \eta_{ir_b'}, \\
M_f(t) &= \sum_i t_i \eta_{ir_f} \eta_{ir_f'} = -M_f(t)^\tau, \\
M_{fb}(t) &= \sum_i t_i \eta_{ir_b} \eta_{ir_f}, \\
N_b(\theta) &= \sum_{i_f} \theta_{2i_f-1} \theta_{2i_f} \eta_{i_f r_b}, \\
N_f(\theta) &= \sum_{i_f} \theta_{2i_f-1} \theta_{2i_f} \eta_{i_f r_f}.
\end{aligned} \tag{9.24}$$

Since the η_{i,r_f} are of degree one and the η_{i,r_b} of degree zero, the matrices M_b and M_f are of degree zero, the M_{bf} and M_{fb} of degree one, while the N_b and N_f are, respectively, of degree zero and one. Thus, the expression (9.23) is of degree zero. Notice that, since the η_{ir_f} are of order one, the matrix $M_f(t)$ is antisymmetric. We also set $M_{bf}(t) = M_{fb}(t) = M_{fb}(t)^\tau$.

We then consider an integral of the form

$$\int \frac{d^D\phi_1 \cdots d^D\phi_{\ell_b}\, d^D\sigma_1 \cdots d^D\sigma_{\ell_f}}{(\sum_i t_i p_i^2 + \sum_{i_f} \theta_{2i_f-1}\theta_{2i_f}\phi_{i_f})^{n-f}} =$$

$$\int \frac{d^D\phi_1 \cdots d^D\phi_{\ell_b}\, d^D\sigma_1 \cdots d^D\sigma_{\ell_f}}{\mathcal{R}(\phi,\sigma,\theta,t)^{n-f}}, \tag{9.25}$$

where

$$\mathcal{R}(\phi,\sigma,\theta,t) = T + \phi^\tau M_b(t)\phi + N_b(\theta)\phi - \sigma^\tau M_f(t)\sigma$$
$$+ \sigma^\tau M_{fb}(t)\phi - \phi^\tau M_{fb}(t)^\tau\sigma + N_f(\theta)\sigma \tag{9.26}$$

and where the integrations in the variables $d^D\sigma_i = d\sigma_{i1}\cdots d\sigma_{iD}$ are Grassmann variables integrations satisfying the properties of the Berezinian integral (9.10) and (9.11), while the integrations in $d^D\phi_i$ are ordinary integrations.

Recall that for Grassmann variables we have the following change of variable formula. Suppose given an invertible antisymmetric $N \times N$ matrix A with entries of degree zero and an N-vector J with entries of degree one. Then we have

$$\sigma^\tau A\sigma + \frac{1}{2}(J^\tau\sigma - \sigma^\tau J) = \eta^\tau A\eta + \frac{1}{4}J^\tau A^{-1}J, \tag{9.27}$$

for $\eta = \sigma - \frac{1}{2}A^{-1}J$. This follows immediately since $A^\tau = -A$ and we have

$$\eta^\tau A\eta = \sigma^\tau A\sigma + \frac{1}{2}J^\tau\sigma - \frac{1}{2}\sigma^\tau J - \frac{1}{4}J^\tau A^{-1}J.$$

We then use this change of variable to write

$$-\sigma^\tau M_f(t)\sigma + \sigma^\tau M_{fb}(t)\phi - \phi^\tau M_{fb}(t)^\tau\sigma + \frac{1}{2}(\sigma^\tau N_f(\theta) - N_f(\theta)^\tau\sigma) =$$

$$-\eta^\tau M_f(t)\eta - \frac{1}{4}(M_{fb}(t)\phi + \frac{1}{2}N_f(\theta))^\tau M_f(t)^{-1}(M_{fb}(t)\phi + \frac{1}{2}N_f(\theta))$$
$$\tag{9.28}$$

with

$$\eta = \sigma - \frac{1}{2}M_f(t)^{-1}\left(M_{fb}(t)\phi + \frac{1}{2}N_f(\theta)\right). \tag{9.29}$$

We have

$$\frac{1}{4}(M_{fb}(t)\phi + \frac{1}{2}N_f(\theta))^\tau M_f(t)^{-1}(M_{fb}(t)\phi + \frac{1}{2}N_f(\theta)) =$$

$$\frac{1}{4}\phi^\tau M_{bf}(t)M_f(t)^{-1}M_{fb}(t)\phi$$

$$+\frac{1}{8}(N_f(\theta)^\tau M_f(t)^{-1}M_{fb}(t)\phi + \phi^\tau M_{bf}(t)M_f(t)^{-1}N_f(\theta))$$

$$+\frac{1}{16}N_f(\theta)^\tau M_f(t)^{-1}N_f(\theta).$$

We then let

$$U(t,\theta,\phi) := T + C(t,\theta) + \phi^\tau A_b(t)\phi + B_b(t,\theta)\phi, \qquad (9.30)$$

where

$$A_b(t) = M_b(t) - \tfrac{1}{4}M_{bf}(t)M_f(t)^{-1}M_{fb}(t)$$

$$B_b(t,\theta) = N_b(\theta) - \tfrac{1}{4}N_f(\theta)^\tau M_f(t)^{-1}M_{fb}(t) \qquad (9.31)$$

$$C(t,\theta) = -\tfrac{1}{16}N_f(\theta)^\tau M_f(t)^{-1}N_f(\theta).$$

Thus, we write the denominator of (9.25) in the form

$$U(t,\theta,\phi)^{n-f}\left(1 + \frac{1}{2}\eta^\tau X_f(t,\theta,\phi)\eta\right)^{n-f}, \qquad (9.32)$$

where we use the notation

$$X_f(t,\theta,\phi) := 2U(t,\theta,\phi)^{-1}M_f(t). \qquad (9.33)$$

Thus, the Grassmann integration in (9.25) gives, as in Lemma 9.2.1,

$$\int \frac{d^D\eta_1 \cdots d^D\eta_{\ell_f}}{\left(1 + \frac{1}{2}\eta^\tau X_f(t,\theta,\phi)\eta\right)^{n-f}} = C_{n,f,\ell_f}\frac{2^{D\ell_f/2}}{U(t,\theta,\phi)^{D\ell_f/2}}\det(M_f(t))^{D/2}, \qquad (9.34)$$

where C_{n,f,ℓ_f} is a combinatorial factor, as we have already encountered in the case of ordinary parametric Feynman integrals for ordinary variables.

We then proceed to the remaining ordinary integration in (9.25). We have, dropping a multiplicative constant,

$$\det(M_f(t))^{D/2}\int \frac{d^D\phi_1 \cdots d^D\phi_{\ell_b}}{U(t,\theta,\phi)^{n-f+D\ell_f/2}}. \qquad (9.35)$$

We use the change of variables $\psi = \phi + \frac{1}{2}M_b(t)^{-1}N_b(\theta)^\tau$. We then have

$$\psi^\tau A_b(t)\psi = \phi^\tau A_b(t)\phi + \tfrac{1}{2}\phi^\tau B_b(t,\theta)^\tau$$
$$+\tfrac{1}{2}B_b(t,\theta)\phi + \tfrac{1}{4}B_b(t,\theta)A_b(t)^{-1}B_b(t,\theta)^\tau, \qquad (9.36)$$

where $A_b(t)^\tau = A_b(t)$ and $(B_b(t,\theta)\phi)^\tau = B_b(t,\theta)\phi$.

We then rewrite (9.35) in the form

$$\det(M_f(t))^{D/2}\int \frac{d^D\psi_1 \cdots d^D\psi_{\ell_b}}{(T + C - \tfrac{1}{4}B_b A_b^{-1}B_b^\tau + \psi^\tau A_b\psi)^{n-f+D\ell_f/2}}. \qquad (9.37)$$

Set then

$$\tilde{T}(t,\theta) = T(t,\theta) + C(t,\theta) - \frac{1}{4}B_b(t,\theta)A_b^{-1}(t)B_b(t,\theta)^\tau, \qquad (9.38)$$

so that we write the above as

$$\frac{\det(M_f(t))^{D/2}}{\tilde{T}(t,\theta)^{n-f+D\ell_f/2}} \int \frac{d^D\psi_1 \cdots d^D\psi_{\ell_b}}{(1 + \psi^\tau X_b(t,\theta)\psi)^{n-f+D\ell_f/2}},$$

with

$$X_b(t,\theta) = \tilde{T}(t,\theta)^{-1}A_b(t).$$

Then, up to a multiplicative constant, the integral gives

$$\tilde{T}^{-n+f-\frac{D\ell_f}{2}+\frac{D\ell_b}{2}} \frac{\det(M_f(t))^{D/2}}{\det(A_b(t))^{D/2}}. \qquad (9.39)$$

Consider first the term

$$\frac{\det(M_f(t))^{D/2}}{\det(A_b(t))^{D/2}}$$

in (9.39) above. This can be identified with a Berezinian. In fact, we have

$$\frac{\det(M_f(t))^{D/2}}{\det(M_b(t) - \frac{1}{4}M_{fb}(t)M_f(t)^{-1}M_{fb}(t))^{D/2}} = \mathrm{Ber}(\mathcal{M}(t))^{-D/2}, \qquad (9.40)$$

where

$$\mathcal{M}(t) = \begin{pmatrix} M_b(t) & \frac{1}{2}M_{fb}(t) \\ \frac{1}{2}M_{bf}(t) & M_f(t) \end{pmatrix}. \qquad (9.41)$$

We now look more closely at the remaining term $\tilde{T}^{-n+f-\frac{D\ell_f}{2}+\frac{D\ell_b}{2}}$ in (9.39). We know from (9.38), (9.31), and (9.24) that we can write $\tilde{T}(t,\theta)$ in the form

$$\tilde{T}(t,\theta) = \sum_i u_i^2 t_i + \sum_j \psi_i \theta_{2j-1}\theta_{2j} + \sum_{i<j} C_{ij}(t)\theta_{2i-1}\theta_{2i}\theta_{2j-1}\theta_{2j}, \qquad (9.42)$$

where the first sum is over all edges and the other two sums are over fermionic edges. We set $\lambda_i = \theta_{2i-1}\theta_{2i}$. Using a change of variables

$$\tilde{\lambda}_i = \lambda_i + \frac{1}{2}C\psi,$$

we rewrite the above as

$$\tilde{T}(t,\theta) = \sum_i u_i^2 t_i - \frac{1}{4}\psi^\tau C\psi + \sum_{i<j} C_{ij}\eta_{2i-1}\eta_{2i}\eta_{2j-1}\eta_{2j},$$

with

$$\tilde{\lambda}_i = \eta_{2i-1}\eta_{2i}.$$

We write

$$\hat{T}(t) = \sum_i u_i^2 t_i - \frac{1}{4}\psi^\tau C\psi$$

and we write

$$\tilde{T}^{-\alpha} = \hat{T}^{-\alpha} \sum_{k=0}^{\infty} \binom{-\alpha}{k} \left(\frac{\frac{1}{2}\tilde{\lambda}^\tau C\tilde{\lambda}}{\hat{T}}\right)^k,$$

where we use the notation

$$\frac{1}{2}\tilde{\lambda}^\tau C\tilde{\lambda} = \sum_{i<j} C_{ij}\eta_{2i-1}\eta_{2i}\eta_{2j-1}\eta_{2j}.$$

Thus, we can write the Feynman integral in the form

$$\int \frac{q_1 \cdots q_f}{q_1 \cdots q_n} d^D s_1 \cdots d^D s_{\ell_b} d^D \sigma_1 \cdots d^D \sigma_{\ell_f} =$$

$$\kappa \int_{\Sigma_{n|2f}} \frac{\Lambda(t)\eta_1 \cdots \eta_{2f}}{\hat{T}(t)^{n - \frac{f}{2} + \frac{D}{2}(\ell_f - \ell_b)} \mathrm{Ber}(\mathcal{M}(t))^{D/2}} dt_1 \cdots dt_n \, d\eta_1 \cdots d\eta_{2f}, \quad (9.43)$$

where $\Lambda(t)$ is $\hat{T}^{f/2}$ times the coefficient of $\eta_1 \cdots \eta_{2f}$ in the expansion

$$\sum_{k=0}^{\infty} \binom{-\alpha}{k} \left(\frac{\frac{1}{2}\tilde{\lambda}^\tau C\tilde{\lambda}}{\hat{T}}\right)^k.$$

More explicitly, this term is of the form

$$\Lambda(t) = \sum C_{i_1 i_2}(t) \cdots C_{i_{f-1} i_f}(t),$$

over indices i_a with $i_{2a-1} < i_{2a}$ and for $k = f/2$. The multiplicative constant in front of the integral on the right hand side above is given by

$$\kappa = \binom{-n + f - \frac{D}{2}(\ell_f - \ell_b)}{f/2}.$$

After imposing $n - \frac{f}{2} + \frac{D}{2}(\ell_f - \ell_b) = 0$ and performing the Grassmann integration of the resulting term

$$\int_{\sigma_{n|2f}} \frac{\Lambda(t)\eta_1 \cdots \eta_{2f}}{\mathrm{Ber}(\mathcal{M}(t))^{D/2}} dt_1 \cdots dt_n \, d\eta_1 \cdots d\eta_{2f}, \quad (9.44)$$

the result follows directly from (9.43). $\qquad\qquad\qquad\qquad\qquad\square$

9.3 Graph supermanifolds

The result of the previous section shows that we have an analog of the period integral

$$\int_{\sigma_n} \frac{dt_1 \cdots dt_n}{\det(M_\Gamma(t))^{D/2}}$$

given by the similar expression

$$\int_{\sigma_n} \frac{\Lambda(t)}{\mathrm{Ber}(\mathcal{M}(t))^{D/2}} dt_1 \cdots dt_n. \tag{9.45}$$

Again we see that, in this case, divergences arise from the intersections between the domain of integration given by the simplex σ_n and the variety defined by the solutions of the equation

$$\frac{\mathrm{Ber}(\mathcal{M}(t))^{D/2}}{\Lambda(t)} = 0. \tag{9.46}$$

Lemma 9.3.1. *For generic graphs, the set of zeros of (9.46) defines a hypersurface in $\mathbb{P}^n$, hence a divisor in $\mathbb{P}^{n-1|2f}$ of dimension $(n-2|2f)$. The support of this divisor is the same as that of the principal divisor defined by $\mathrm{Ber}(\mathcal{M}(t))$.*

Proof. The generic condition on graphs is imposed to avoid the cases with $M_f(t) \equiv 0$. Thus, suppose given a pair (Γ, B) that is generic, in the sense that $M_f(t)$ is not identically zero. The equation (9.46) is satisfied by solutions of

$$\det(M_b(t) - \frac{1}{4} M_{bf}(t) M_f(t)^{-1} M_{fb}(t)) = 0 \tag{9.47}$$

and by poles of $\Lambda(t)$. (The factor $1/4$ in (9.47) comes from the change of variable formula (9.27).) Using the formulae (9.31) and (9.24) we see that the denominator of $\Lambda(t)$ is given by powers of $\det(M_f(t))$ and $\det(A_b(t)) = \det(M_b(t) - \frac{1}{4} M_{bf}(t) M_f(t)^{-1} M_{fb}(t))$. Thus, the set of solutions of (9.46) is the union of zeros and poles of $\mathrm{Ber}(\mathcal{M}(t))$. The multiplicities are given by the powers of these determinants that appear in $\Lambda(t) \mathrm{Ber}(\mathcal{M}(t))^{-D/2}$. $\square$

Definition 9.3.2. Let Γ be a graph with bosonic and fermionic edges and B a choice of a basis of $H_1(\Gamma)$. We denote by $\mathcal{X}_{(\Gamma,B)} \subset \mathbb{P}^{n-1|2f}$ the locus of zeros and poles of $\mathrm{Ber}(\mathcal{M}(t)) = 0$. We refer to $\mathcal{X}_{(\Gamma,B)}$ as the *graph supermanifold*.

In the degenerate cases of graphs such that $M_f(t) \equiv 0$, we simply set $\mathcal{X}_{(\Gamma,B)} = \mathbb{P}^{n-1|2f}$. Examples of this sort are provided by data (Γ, B) such that there is only one loop in B containing fermionic edges. Other special cases arise when we consider graphs with only bosonic or only fermionic edges. In the first case, we go back to the original calculation without Grassmann variables and we therefore simply recover

$$\mathcal{X}_{(\Gamma,B)} = X_\Gamma = \{t : \det(M_b(t)) = 0\} \subset \mathbb{P}^{n-1|0}.$$

In the case with only fermionic edges, we have $\det(M_b(t) - \frac{1}{4}M_{bf}(t)M_f(t)^{-1}M_{fb}(t)) \equiv 0$ since both $M_b(t)$ and $M_{bf}(t)$ are identically zero. It is then natural to simply assume that, in such cases, the graph supermanifold is simply given by $\mathcal{X}_{(\Gamma,B)} = \mathbb{P}^{f-1|2f}$.

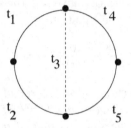

Consider the example of the figure, with a choice of the basis of loops satisfying the condition (9.18). It is easy to see that such a choice is possible: it has two generators, one of them a loop made of fermionic edges and the second a loop containing both fermionic and bosonic edges. Let us assign the ordinary variables t_i with $i = 1, \ldots, 5$ to the edges as in the figure above. We then have

$$M_b(t) = t_1 + t_2 + t_3$$

since only the second loop in the basis contains bosonic edges, while we have

$$M_{bf}(t) = (t_1 + t_2, t_1 + t_2 + t_3)$$

$$= t_1(1,1) + t_2(1,1) + t_3(0,1) + t_4(0,0) + t_5(0,0)$$

and

$$M_f(t) = \begin{pmatrix} 0 & t_1 + t_2 \\ -(t_1 + t_2) & 0 \end{pmatrix}.$$

Thus, we obtain in this case

$$M_{bf}(t)M_f(t)^{-1}M_{fb}(t) =$$

$$(t_1 + t_2, t_1 + t_2 + t_3) \begin{pmatrix} 0 & \frac{-1}{t_1+t_2} \\ \frac{1}{t_1+t_2} & 0 \end{pmatrix} \begin{pmatrix} t_1 + t_2 \\ t_1 + t_2 + t_3 \end{pmatrix}$$

$$= (t_1 + t_2, t_1 + t_2 + t_3) \begin{pmatrix} \frac{-(t_1+t_2+t_3)}{t_1+t_2} \\ 1 \end{pmatrix}$$

$$= -(t_1 + t_2 + t_3) + t_1 + t_2 + t_3 \equiv 0.$$

Thus, in this particular example we have $M_{bf}(t)M_f(t)^{-1}M_{fb}(t) \equiv 0$ for all $t = (t_1, \dots, t_5)$, so that

$$\mathrm{Ber}(\mathcal{M}(t)) = \det(M_b(t)) \det(M_f(t))^{-1} = (t_1 + t_2 + t_3)/(t_1 + t_2)^2$$

and the locus of zeros and poles $\mathcal{X}_{(\Gamma, B)} \subset \mathbb{P}^{5|8}$ is the union of $t_1 + t_2 + t_3 = 0$ and $t_1 + t_2 = 0$ in $\mathbb{P}^5$ (the latter counted with multiplicity two), with the restriction of the sheaf from $\mathbb{P}^{5|8}$.

It is also shown in [Marcolli and Rej (2008)] that the universality result of [Belkale and Brosnan (2003a)] implies a similar result for the supermanifold case, using the simple relation of Proposition 9.1.2 between the Grothendieck ring of ordinary manifolds and of supermanifolds.

9.4 Noncommutative field theories

Scalar quantum field theories, especially ϕ^4 theories, over a noncommutative spacetime have been investigated extensively in recent years, especially in relation to their renormalization properties. Such theories arise as effective limits of string theory [Connes, Douglas, Schwarz (1998)], [Seiberg and Witten (1999)], where the underlying spacetime is deformed to a Moyal type noncommutative space.

We do not give here a detailed exposition of these theories, and we refer the reader to the lectures [Grosse and Wulkenhaar (2008)] for a nice and detailed overview. We only recall that in dimension $D = 4$, when the underlying $\mathbb{R}^4$ is made noncommutative by deformation to $\mathbb{R}^4_\theta$ with the Moyal product, the theory suffers from a new type of divergences, coming from UV/IR (ultraviolet/infrared) mixing. These render the theory nonrenormalizable. However, perturbative renormalization can be restored in the ϕ^4 theory in different ways. In the Grosse–Wulkenhaar model [Grosse and Wulkenhaar (2005)], this is done by the addition of a harmonic oscillator term that is quite natural from the point of view of the Moyal product, but which breaks the translation invariance of the theory. More recently, a different approach to restoring renormalization for the ϕ^4 theory on Moyal spaces, which maintains translation invariance, was introduced in [Gurau, Magnen, Rivasseu, Tanasa (2009)]. Both models are renormalizable to all orders.

What interests us most here is the fact that, in these theories, it is also possible to obtain a parametric representation of Feynman integrals. Due to the fact that the underlying spacetime coordinates no longer commute, ordinary Feynman graphs are replaced in this context by ribbon graphs and the planarity or non-planarity of the ribbon graphs plays an important role in the properties of the corresponding Feynman integrals. We concentrate on some special examples among the parametric representation described in [Gurau and Rivasseu (2007)], [Tanasa (2007)].

Among the family of "banana graphs", whose motivic properties were described in [Aluffi and Marcolli (2008a)], the only ones that can appear as Feynman graphs of a ϕ^4 theory are the one-loop case (with two external edges at each vertex), the two-loop case (with one external edge at each vertex) and the three-loop case as a vacuum bubble. Excluding the vacuum bubble because of the presence of the polynomial $P_\Gamma(t)$, we see that the effect of making the underlying spacetime noncommutative turns the remaining two graphs into the following ribbon graphs. (Notice how the two-loop ribbon graph now has two distinct versions, only one of which is a planar graph.)

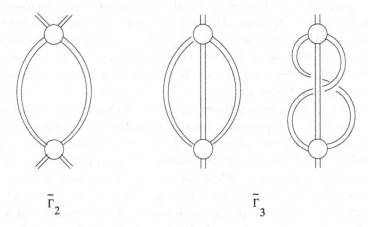

$$\tilde{\Gamma}_2 \qquad\qquad \tilde{\Gamma}_3$$

The usual Kirchhoff polynomial $\Psi_\Gamma(t)$ of the Feynman graph, as well as the polynomial $P_\Gamma(t,p)$, are replaced by new polynomials involving pairs of spanning trees, one in the graph itself and one in another associated graph which is a dual graph in the planar case and that is obtained from an embedding of the ribbon graph on a Riemann surface in the more general case. Unlike the commutative case, these polynomials are *no longer homogeneous*, hence the corresponding graph hypersurface only makes sense as an affine hypersurface. The relation of the hypersurface for the noncommutative case and the one of the original commutative case (also viewed as an affine hypersurface) is given by the following statement.

Proposition 9.4.1. *Let $\tilde{\Gamma}$ be a ribbon graph in the noncommutative ϕ^4-theory that corresponds in the ordinary ϕ^4-theory to a graph Γ with n internal edges. Then instead of a single graph hypersurface $\hat{X}_\Gamma$ one has a one-parameter family of affine hypersurfaces $\hat{X}_{\tilde{\Gamma},s} \subset \mathbb{A}^n$, where the parameter $s \in \mathbb{R}_+$ depends upon the deformation parameter θ of the noncommutative $\mathbb{R}^4_\theta$ and on the parameter Ω of the harmonic oscillator term in the Grosse–Wulkenhaar model. The hypersurface corresponding to the value $s = 0$ has a singularity at the origin $0 \in \mathbb{A}^n$ whose tangent cone is the (affine) graph hypersurface $\hat{X}_\Gamma$.*

Proof. This follows directly from the relation between the graph polynomial for the ribbon graph $\tilde{\Gamma}$ given in [Gurau and Rivasseu (2007)] and the Kirchhoff polynomial Ψ_Γ. It suffices to see that (a constant multiple of) the Kirchhoff polynomial is contained in the polynomial for $\tilde{\Gamma}$ for all values of the parameter s, and that it gives the part of lowest order in the variables t_i when $s = 0$. □

In the specific examples of the banana graphs $\tilde{\Gamma}_2$ and $\tilde{\Gamma}_3$ of the figure above, the polynomials have been computed explicitly in [Gurau and Rivasseu (2007)] and they are of the form

$$\Psi_{\tilde{\Gamma}_2} = (1 + 4s^2)(t_1 + t_2 + t_1^2 t_2 + t_1 t_2^2), \tag{9.48}$$

where the parameter $s = (4\theta\Omega)^{-1}$ is a function of the deformation parameter $\theta \in \mathbb{R}$ of the Moyal plane and of the parameter Ω in the harmonic oscillator term in the Grosse–Wulkenhaar action functional (see [Grosse and Wulkenhaar (2008)]). One can see the polynomial $\Psi_{\Gamma_2}(t) = t_1 + t_2$ appearing as lowest order term. Similarly for the two graphs $\tilde{\Gamma}_3$ that correspond to the banana graph Γ_3 one has ([Gurau and Rivasseu (2007)])

$$\Psi_{\tilde{\Gamma}_3}(t) = (1 + 8s^2 + 16s^4)(t_1 t_2 + t_2 t_3 + t_1 t_3 + t_1^2 t_2 t_3 + t_1 t_2^2 t_3 + t_1 t_2 t_3^2)$$
$$+ 16s^2(t_2^2 + t_1^2 t_3^2)$$

$$\tag{9.49}$$

for the planar case, while for the non-planar case one has

$$\Psi_{\tilde{\Gamma}_3}(t) = (1 + 8s^2 + 16s^4)(t_1 t_2 + t_2 t_3 + t_1 t_3 + t_1^2 t_2 t_3 + t_1 t_2^2 t_3 + t_1 t_2 t_3^2)$$
$$+ 4s^2(t_2^2 + t_1^2 t_3^2 + t_1^2 + t_2^2 t_3^2 + t_3^2 + t_1^2 t_2^2 + 1 + t_1^2 t_2^2 t_3^2).$$
(9.50)

In both cases, one readily recognizes the polynomial $\Psi_{\Gamma_3}(t) = t_1 t_2 + t_2 t_3 + t_1 t_3$ as the lowest order part at $s = 0$. Notice how, when $s \neq 0$, one finds other terms of order less than or equal to that of the polynomial $\Psi_{\Gamma_3}(t)$, such as t_2^2 in (9.49) and $1 + t_1^2 + t_2^2 + t_3^2$ in (9.50). Notice also how, at the limit value $s = 0$ of the parameter, the two polynomials for the two different ribbon graphs corresponding to the third banana graph Γ_3 agree.

For each value of the parameter $s = (4\theta\Omega)^{-1}$ one obtains in this way an affine hypersurface, which is a curve in $\mathbb{A}^2$ or a surface in $\mathbb{A}^3$, and that has the corresponding affine $\hat{X}_{\Gamma_n}$ as tangent cone at the origin in the case $s = 0$. The latter is a line in the $n = 2$ case and a cone on a nonsingular conic in the case $n = 3$.

Following [Aluffi and Marcolli (2008a)], we now look at what happens to the invariants such as the Chern–Schwartz–MacPherson (CSM) class of the graph hypersurfaces, when passing to the nocommutative case. An interesting observation one obtains in this way is that the CSM classes detect certain special values of the deformation parameter $s = (4\theta\Omega)^{-1}$ where the hypersurface $\hat{X}_{\tilde{\Gamma}_3}$ becomes more singular and gives a quantitative estimate of the amount by which the singularities change.

The CSM class is naturally defined for projective varieties. In the case of an affine hypersurface defined by a non-homogeneous equation, one can choose to compactify it in projective space by adding an extra variable and making the equation homogeneous and then computing the CSM class of the corresponding projective hypersurface. However, in doing so one should be aware of the fact that the CSM class of an affine variety, defined by choosing an embedding in a larger compact ambient variety, depends on the choice of the embedding. An intrinsic definition of CSM classes for non-compact varieties which does not depend on the embedding was given in [Aluffi (2006)]. However, for our purposes here it suffices to take the simpler viewpoint of making the equation homogeneous and then computing CSM classes with the Macaulay2 program of [Aluffi (2003)].

Proposition 9.4.2. *Let* $\hat{X}_{\tilde{\Gamma}_3}$ *denote the affine surface defined by the equation* (9.49) *and let* $\bar{X}_{\tilde{\Gamma}_3} \subset \mathbb{P}^3$ *be the hypersurface obtained by making the equation* (9.49) *homogeneous. For general values of the parameter*

$s = (4\theta\Omega)^{-1}$ *the CSM class is given by*

$$c(\bar{X}_{\tilde{\Gamma}_3}) = 14H^3 + 4H. \tag{9.51}$$

For the special value $s = 1/2$ of the parameter, the CSM class becomes of the form

$$c(\bar{X}_{\tilde{\Gamma}_3})|_{s=1/2} = 5H^3 + 5H^2 + 4H, \tag{9.52}$$

while in the limit $s \to 0$ one has

$$c(\bar{X}_{\tilde{\Gamma}_3})|_{s=0} = 11H^3 + 4H. \tag{9.53}$$

It is also interesting to notice that, when we consider the second equation (9.50) for the non-planar ribbon graph associated to the third banana graph Γ_3, we see an example where the graph hypersurfaces of the non-planar graphs of noncommutative field theory no longer satisfy the positivity property conjectured for the the the graph hypersurfaces of the commutative field theories in [Aluffi and Marcolli (2008a)]. In fact, as in the case of the equation for the planar graph (9.49), we now find the following result.

Proposition 9.4.3. *Let $X_{\tilde{\Gamma}_3} \subset \mathbb{P}^3$ denote the affine surface defined by the equation (9.50) and let $\bar{X}_{\tilde{\Gamma}_3} \subset \mathbb{P}^3$ be the hypersurface obtained by making the equation (9.50) homogeneous. For general values of the parameter $s = (4\theta\Omega)^{-1}$ the CSM class is given by*

$$c(\bar{X}_{\tilde{\Gamma}_3}) = 33H^3 - 9H^2 + 6H. \tag{9.54}$$

The special case $s = 1/2$ is given by

$$c(\bar{X}_{\tilde{\Gamma}_3})|_{s=1/2} = 9H^3 - 3H^2 + 6H \tag{9.55}$$

Notice that, in the case of ordinary Feynman graphs of commutative scalar field theories, all the examples where the CSM classes of the corresponding hypersurfaces have been computed explicitly (either theoretically or numerically) are planar graphs. Although it seems unlikely that planarity will play a role in the conjectured positivity of the coefficients of the CSM classes in the ordinary case, the example above showing that CSM classes of graph hypersurfaces of non-planar ribbon graphs in noncommutative field theories can have negative coefficients makes it more interesting to check the case of non-planar graphs in the ordinary case as well. It is well known that, for an ordinary graph to be non-planar, it or a quotient by a subgraph has to contain a copy of either the complete graph K_5 on five vertices or the complete bipartite graph $K_{3,3}$ on six vertices. Either one of these graphs corresponds to a graph polynomial that is currently beyond the reach of the available Macaulay2 routine and a theoretical argument that provides a more direct approach to the computation of the corresponding CSM class does not seem to be easily available. It remains an interesting question to compute these CSM classes, especially in view of the fact that the original computations of [Broadhurst and Kreimer (1997)] of Feynman integrals of graphs appear to indicate that the non-planarity of the graph is somehow related to the presence of more interesting periods (*e.g.* multiple as opposed to simple zeta values). It would be interesting to see whether it also has an effect on invariants such as the CSM class.

Bibliography

S. Agarwala, *The Geometry of Renormalization*. PhD Thesis, Johns Hopkins University, 2009.

P. Aluffi, *Limits of Chow groups, and a new construction of Chern-Schwartz-MacPherson classes*. Pure Appl. Math. Q. 2 (2006), no. 4, part 2, 915–941.

P. Aluffi, *Computing characteristic classes of projective schemes*. J. Symbolic Comput. 35 (2003), no. 1, 3–19.

P. Aluffi, M. Marcolli, *Feynman motives of banana graphs*, Communications in Number Theory and Physics, Vol.3 (2009) N.1, 1–57.

P. Aluffi, M. Marcolli, *Algebro-geometric Feynman rules*, arXiv:0811.2514.

P. Aluffi, M. Marcolli, *Parametric Feynman integrals and determinant hypersurfaces*, preprint arXiv:0901.2107, to appear in Advances in Theoretical and Mathematical Physics.

P. Aluffi, M. Marcolli, *Feynman motives and deletion-contraction relations*, arXiv:0907.3225.

P. Aluffi, L.C. Mihalcea, *Chern classes of Schubert cells and varieties*, arXiv:math/0607752, to appear in Journal of Algebraic Geometry.

Y. André, *Une introduction aux motifs*, Societé Mathématique de France, 2004.

Y. André, *An introduction to motivic zeta functions of motives*, arXiv:0812.3920.

D. Arapura, *A category of motivic sheaves*, preprint arXiv:0801.0261.

V.I. Arnold, V.V. Goryunov, O.V. Lyashko, V.A. Vasilev, *Singularity theory, I*, Springer Verlag, 1998.

V.I. Arnold, S.M. Gusein-Zade, A.N. Varchenko, *Singularities of differentiable maps*, Vol.II, Birkhäuser, 1988.

J. Ayoub, *The motivic vanishing cycles and the conservation conjecture*, in "Algebraic Cycles and Motives", London Mathematical Society Lecture Note Series, Vol.343, pp.3–54, Cambridge University Press, 2007.

A. Beĭlinson, J. Bernstein, P. Deligne, *Faisceaux pervers*, in "Analysis and topology on singular spaces, I" (Luminy, 1981), 5–171, Astérisque, 100, Soc. Math. France, Paris, 1982.

A. Beĭlinson, P. Deligne, *Interprétation motivique de la conjecture de Zagier reliant polylogarithmes et régulateurs*. in "Motives" (Seattle, WA, 1991), 97–121, Proc. Sympos. Pure Math., 55, Part 2, Amer. Math. Soc., Providence,

　　　　　　　　　　　　Feynman Motives

RI, 1994.

P. Belkale, P. Brosnan, *Matroids, motives, and a conjecture of Kontsevich*, Duke Math. Journal, Vol.116 (2003) 147–188.

P. Belkale, P. Brosnan, *Periods and Igusa local zeta functions*. Int. Math. Res. Not. 2003, no. 49, 2655–2670.

C. Bergbauer, A. Rej, *Insertion of graphs and singularities of graph hypersurfaces*, preprint 2009.

I.N. Bernšteĭn, D.A. Leĭtes, *Integral forms and the Stokes formula on supermanifolds*. Functional Anal. Appl. 11 (1977), no. 1, 45–47.

J. Bjorken, S. Drell, *Relativistic Quantum Mechanics*, McGraw-Hill, 1964.

J. Bjorken, S. Drell, *Relativistic Quantum Fields*, McGraw-Hill, 1965.

S. Bloch, *Lectures on mixed motives*. Algebraic geometry—Santa Cruz 1995, 329–359, Proc. Sympos. Pure Math., 62, Part 1, Amer. Math. Soc., 1997.

S. Bloch, *Motives associated to graphs*, Japan J. Math., Vol.2 (2007) 165–196.

S. Bloch, *Motives associated to sums of graphs*, arXiv:0810.1313.

S. Bloch, E. Esnault, D. Kreimer, *On motives associated to graph polynomials*, Commun. Math. Phys., Vol.267 (2006) 181–225.

S. Bloch, D. Kreimer, *Mixed Hodge structures and renormalization in physics*, Commun. Number Theory Phys. 2 (2008), no. 4, 637–718.

N.N. Bogoliubov, O. Parasiuk, *Über die Multiplikation der Kausal Funktionen in der Quantentheorie der Felder*, Acta Mathematica 97 (1957) 227–266.

C. Bogner, S. Weinzierl, *Periods and Feynman integrals*, J. Math. Phys. 50 (2009), no. 4, 042302, 16 pp.

C. Bogner, S. Weinzierl, *Resolution of singularities for multi-loop integrals*, arXiv:0709.4092.

J.B. Bost, A. Connes, *Hecke algebras, Type III factors and phase transitions with spontaneous symmetry breaking in number theory*, Selecta Math. (New Series) Vol.1 (1995) N.3, 411–457.

P. Breitenlohner, D. Maison, *Dimensional renormalization and the action principle*, Comm. Math. Phys. Vol.52 (1977) N.1, 11–38.

T. Bridgeland, *Spaces of stability conditions*, Proc. Symposium Pure Math., Vol. 80 (2009) N.1, 1–21.

D. Broadhurst, D. Kreimer, *Association of multiple zeta values with positive knots via Feynman diagrams up to 9 loops*, Phys. Lett. B, Vol.393 (1997) 403–412.

F. Brown, *The massless higher-loop two-point function*, Commun. Math. Phys. Vol.287 (2009) 925–958.

F. Brown, *On the periods of some Feynman integrals*, arXiv:0910.0114.

F. Brown, *Multiple zeta values and periods of moduli spaces* $\mathfrak{M}_{0,n}$, arXiv:math/0606419.

A. Bruguières, *On a Tannakian theorem due to Nori*, preprint 2004.

A.L. Carey, J. Phillips, A. Rennie, F. A. Sukochev *The Local Index Formula in Semifinite von Neumann Algebras I: Spectral Flow* Adv. Math. 202 (2006), no. 2, 451–516.

A.L. Carey, J. Phillips, A. Rennie, F. A. Sukochev *The Local Index Formula in Semifinite von Neumann Algebras II: The Even Case* Adv. Math. 202 (2006), no. 2, 517–554.

J. Collins, *Renormalization*, Cambridge University Press, 1986.

A. Connes, *Une classification des facteurs de type III*. Ann. Sci. École Norm. Sup. (4) 6 (1973), 133–252.

A. Connes, *Geometry from the spectral point of view*. Lett. Math. Phys. 34 (1995), no. 3, 203–238.

A. Connes, *Trace formula in noncommutative geometry and the zeros of the Riemann zeta function*. Selecta Math. (N.S.) 5 (1999), no. 1, 29–106.

A. Connes, M. Douglas, A. Schwarz, *Noncommutative geometry and matrix theory: compactification on tori*. JHEP 9802 (1998) 3–43.

A. Connes, D. Kreimer, *Renormalization in quantum field theory and the Riemann–Hilbert problem I. The Hopf algebra structure of graphs and the main theorem*, Comm. Math. Phys., Vol.210 (2000) 249–273.

A. Connes, D. Kreimer, *Renormalization in quantum field theory and the Riemann-Hilbert problem. II. The β-function, diffeomorphisms and the renormalization group*. Comm. Math. Phys. 216 (2001), no. 1, 215–241.

A. Connes, M. Marcolli, *Renormalization and motivic Galois theory*. International Math. Res. Notices (2004) N.76, 4073–4091.

A. Connes, M. Marcolli, *Quantum fields and motives*, Journal of Geometry and Physics, Volume 56, (2005) N.1, 55–85.

A. Connes, M. Marcolli, *Anomalies, Dimensional Regularization and noncommutative geometry*, unpublished manuscript, 2005, available at www.its.caltech.edu/~matilde/work.html

A. Connes, M. Marcolli, *From physics to number theory via noncommutative geometry, Part I: Quantum statistical mechanics of $\mathbb{Q}$-lattices*, in "Frontiers in Number Theory, Physics, and Geometry. I", pp.269–347, Springer Verlag, 2006.

A. Connes, M. Marcolli, *From Physics to Number Theory via Noncommutative Geometry. Part II: Renormalization, the Riemann-Hilbert correspondence, and motivic Galois theory*, in "Frontiers in Number Theory, Physics, and Geometry, II" pp.617–713, Springer Verlag, 2006.

A. Connes, M. Marcolli, *Noncommutative Geometry, Quantum Fields, and Motives*, Colloquium Publications, Vol.55, American Mathematical Society, 2008.

A. Connes, H. Moscovici, *The local index formula in noncommutative geometry*, Geom. Funct. Anal. 5 (1995), no. 2, 174–243.

C. Consani, *Double complexes and Euler L-factors*. Compositio Math. 111 (1998), no. 3, 323–358.

C. Consani, M. Marcolli, *Noncommutative geometry, dynamics, and ∞-adic Arakelov geometry*. Selecta Math. (N.S.) 10 (2004), no. 2, 167–251.

C. Consani, M. Marcolli, *Archimedean cohomology revisited*, in "Noncommutative geometry and number theory", pp.109–140, Aspects Math., E37, Vieweg Verlag, 2006.

P. Deligne and A. Dimca, *Filtrations de Hodge et par d'ordre du pôle pour les hypersurfaces singulières*, Ann. Sci. École Norm. Sup. 23 (1990) 645–656.

P. Deligne, A.B. Goncharov, *Groupes fondamentaux motiviques de Tate mixte*, Ann. Sci. École Norm. Sup. (4) 38 (2005), no. 1, 1–56.

P. Deligne, J.S. Milne, *Tannakian categories*, in "Hodge cycles, motives, and Shimura varieties", Lecture Notes in Mathematics, 900, pp.101–228. Springer-Verlag, 1982.

M. Demazure, A. Grothendieck, et al. *Séminaire Géometrie Algébrique: Schémas en Groupes*, Lecture Notes in Mathematics, Vol. 151, 152, 153. Springer, 1970.

J. Denef, F. Loeser, *Motivic Igusa zeta functions*. J. Algebraic Geom. 7 (1998), no. 3, 505–537.

A. Dimca, *Singularities and topology of hypersurfaces*, Universitext, Springer-Verlag, 1992.

I. Dolgachev, *Weighted projective varieties*, in "Group actions and vector fields", LNM 956, pp. 34–71, Springer-Verlag, 1982.

D. Doryn, *Cohomology of graph hypersurfaces associated to certain Feynman graphs*, PhD Thesis, University of Duisburg-Essen, arXiv:0811.0402.

K. Ebrahimi-Fard, L. Guo, D. Kreimer, *Spitzer's identity and the algebraic Birkhoff decomposition in pQFT*. J. Phys. A 37 (2004), no. 45, 11037–11052.

K. Ebrahimi-Fard, J.M. Gracia-Bondía, F. Patras, *A Lie theoretic approach to renormalization*, Commun.Math.Phys. Vol.276 (2007) 519–549.

I.M. Gel'fand, S.G. Gindikin, M.I. Graev, *Integral geometry in affine and projective spaces*. Current problems in mathematics, Vol. 16, pp. 53–226, 228, Akad. Nauk SSSR, Vsesoyuz. Inst. Nauchn. i Tekhn. Informatsii, Moscow, 1980.

S.I. Gelfand, Yu.I. Manin, *Homological algebra*, Encyclopedia of Mathematical Sciences, Vol.38, Springer Verlag, 1994.

I.M. Gelfand, V.V. Serganova, *Combinatorial geometries and the strata of a torus on homogeneous compact manifolds*. Uspekhi Mat. Nauk 42 (1987), no. 2(254), 107–134, 287.

H. Gillet, C. Soulé, *Descent, motives and K-theory*. J. Reine Angew. Math. 478 (1996), 127–176.

A.B. Goncharov, *Volumes of hyperbolic manifolds and mixed Tate motives*. J. Amer. Math. Soc. 12 (1999), no. 2, 569–618.

A.B. Goncharov, *Multiple polylogarithms and mixed Tate motives*, math.AG/0103059.

A.B. Goncharov, Yu.I. Manin, *Multiple ζ-motives and moduli spaces $\overline{\mathcal{M}}_{0,n}$*. Compos. Math. 140 (2004), no. 1, 1–14.

H. Grosse, R. Wulkenhaar, *Renormalization of noncommutative quantum field theory*, in "An invitation to Noncommutative Geometry" pp.129–168, World Scientific, 2008.

H. Grosse, R. Wulkenhaar, *Renormalization of ϕ^4-theory on noncommutative $\mathbb{R}^4$ in the matrix base*, Commun. Math. Phys. 256 (2005) 305–374.

F. Guillén, V. Navarro Aznar, *Sur le théorème local des cycles invariants*. Duke Math. J. 61 (1990), no. 1, 133–155.

R. Gurau, J. Magnen, V. Rivasseau, A. Tanasa, *A translation-invariant renormalizable non-commutative scalar model*, Commun. Math. Phys. 287 (2009) 275–290.

R. Gurau, V. Rivasseau, *Parametric Representation of Noncommutative Field*

Theory, Commun. Math. Phys. Vol. 272 (2007) N.3, 811–835.

M. Hanamura, *Mixed motives and algebraic cycles. I.* Math. Res. Letters, Vol.2 (1995) N.6, 811–821.

K. Hepp, *Proof of the Bogoliubov-Parasiuk theorem on renormalization*, Comm. Math. Phys. 2, (1966), 301–326.

C. Itzykson, J.B. Zuber, *Quantum Field Theory*, Dover Publications, 2006.

U. Jannsen, *Motives, numerical equivalence, and semi-simplicity*, Invent. Math. 107 (1992) N.3, 447–452.

U. Jannsen, S. Kleiman and J.P. Serre (Eds.) *Motives*. Proceedings of Symposia in Pure Mathematics, 55, Part 1-2. American Mathematical Society, Providence, RI, 1994.

F. Jegerlehner, *Facts of life with* γ_5, Eur. Phys. J. C 18 (2001) 673–679.

M. Kapranov, *The elliptic curve in the S-duality theory and Eisenstein series for Kac-Moody groups*, arXiv:math.AG/0001005.

M. Kashiwara, P. Schapira, *Sheaves on manifolds*, Springer, 1990.

D. Kazhdan, G. Lusztig, *Schubert varieties and Poincaré duality*. in "Geometry of the Laplace operator" pp. 185–203, Proc. Sympos. Pure Math., XXXVI, Amer. Math. Soc. 1980.

M. Kontsevich, D. Zagier, *Periods*. in "Mathematics unlimited—2001 and beyond", pp.771–808, Springer, 2001.

D. Kreimer, *The core Hopf algebra*, preprint 2009.

D. Kreimer, W.D. van Suijlekom, *Recursive relations in the core Hopf algebra*, arXiv:0903.2849.

K. Kremnizer, M. Szczesny, *Feynman graphs, rooted trees, and Ringel-Hall algebras*, arXiv:0806.1179.

V.S. Kulikov, *Mixed Hodge structures and singularities*. Cambridge University Press, 1998.

S. Lang, $SL_2(\mathbb{R})$, Addison–Wesley, 1975.

M. Larsen, V. Lunts, *Motivic measures and stable birational geometry*, Moscow Math. Journal, Vol.3 (2003) N.1, 85–95.

M. LeBellac, *Quantum and statistical field theory*, Oxford Science Publications, 1991.

M. Levine, *Tate motives and the vanishing conjectures for algebraic K-theory*, in Algebraic K-theory and algebraic topology (Lake Louise, AB, 1991), 167188, NATO Adv. Sci. Inst. Ser. C Math. Phys. Sci., 407, Kluwer Acad. Publ., 1993.

M. Levine, *Mixed motives*, Math. Surveys and Monographs, Vol.57, American Mathematical Society, 1998.

M. Levine, *Motivic Tubular Neighborhoods*, arXiv:math/0509463

R.D. MacPherson, *Chern classes for singular algebraic varieties*. Ann. of Math. (2) 100 (1974), 423–432.

Yu.I. Manin, *Correspondences, motifs and monoidal transformations*, Math. USSR-Sb. 6 (1968), 439–470.

Yu.I. Manin, *Gauge field theory and complex geometry*, Springer Verlag, 1988.

M. Marcolli, *Motivic renormalization and singularities*, preprint arXiv:0804.4824.

M. Marcolli, A. Rej, *Supermanifolds from Feynman graphs*, Journal of Physics A,

Vol.41 (2008) 315402 (21pp). arXiv:0806.1681.

P.B. Medvedev, α *representation in scalar supersymmetric theory*, Teoretich-eskaya i matematicheskaya fizika (transl.Institute of Theoretical and Experimental Physics) Vol.35 (1978) N.1, 37–47.

B. Mohar, C. Thomassen, *Graphs on Surfaces*, Johns Hopkins University Press, 2001.

N. Mnëv, *The universality theorems on the classification problem of configuration varieties and convex polytopes varieties*, in "Topology and geometry - Rohlin seminar". Lect. Notes Math., vol. 1346, pp. 527–543, Springer 1988.

N. Nakanishi *Graph Theory and Feynman Integrals*. Gordon and Breach, 1971.

J. Rammer, *Quantum field theory of non-equilibrium states*, Cambridge University Press, 2007.

P. Ramond, *Field theory: a modern primer*. Addison Wesley, 1990.

A.A. Rosly, A.S. Schwarz, A.A. Voronov, *Geometry of superconformal manifolds*, Comm. Math. Phys., Vol.119 (1988) 129–152.

N. Saavedra Rivano, *Catégories Tannakiennes*. Lecture Notes in Mathematics, Vol. 265. Springer-Verlag, 1972.

M. Saito, *Modules de Hodge polarisables*. Publ. Res. Inst. Math. Sci. 24 (1988), no. 6, 849–995 (1989).

M. Sato, T. Miwa, M. Jimbo, T. Oshima, *Holonomy structure of Landau singularities and Feynman integrals*, Publ. Res. Inst. Math. Sci. Volume 12, Supplement (1976/1977) (1976), 387–439.

M.H. Schwartz, *Classes caractéristiques définies par une stratification d'une variété analytique complexe. I.* C. R. Acad. Sci. Paris, Vol.260 (1965) 3262–3264.

N. Seiberg, E. Witten, *String theory and noncommutative geometry*, JHEP 9909 (1999) 32–131.

I.R. Shafarevich, *Basic Algebraic Geometry 2: Schemes and Complex Manifolds*. Second edition, Springer, 1996.

J. Steenbrink, *Limits of Hodge structures*. Invent. Math. 31 (1976), 229–257.

J. Steenbrink, *Mixed Hodge structure on the vanishing cohomology*. in "Real and complex singularities" (Proc. Ninth Nordic Summer School/NAVF Sympos. Math., Oslo, 1976), pp. 525–563. Sijthoff and Noordhoff, Alphen aan den Rijn, 1977.

J. Steenbrink, *A summary of mixed Hodge theory*. in "Motives" (Seattle, WA, 1991), 31–41, Proc. Sympos. Pure Math., 55, Part 1, Amer. Math. Soc., Providence, RI, 1994.

J. Stembridge, *Counting points on varieties over finite fields related to a conjecture of Kontsevich*. Ann. Comb. 2 (1998), no. 4, 365–385.

A. Tanasa, *Overview of the parametric representation of renormalizable noncommutative field theory*, arXiv:0709.2270.

B. Teissier, *Introduction to equisingularity problems*. Algebraic geometry (Proc. Sympos. Pure Math., Vol. 29, Humboldt State Univ., Arcata, Calif., 1974), pp. 593–632. Amer. Math. Soc., Providence, R.I., 1975.

T. Terasoma, *Mixed Tate motives and multiple zeta values*. Invent. Math. 149 (2002), no. 2, 339–369.

R. Vakil, *Murphy's law in algebraic geometry: badly-behaved deformation spaces.* Invent. Math. 164 (2006), no. 3, 569–590.

W. van Suijlekom, *Renormalization of gauge fields: a Hopf algebra approach.* Comm. Math. Phys. 276 (2007), no. 3, 773–798.

W. van Suijlekom, *The Hopf algebra of Feynman graphs in quantum electrodynamics.* Lett. Math. Phys. 77 (2006), N.3, 265–281.

W. van Suijlekom, *Representing Feynman graphs on BV-algebras*, preprint, arXiv:0807.0999.

A.N. Varchenko, *Asymptotic behavior of holomorphic forms determines a mixed Hodge structure.* Dokl. Akad. Nauk SSSR 255 (1980), no. 5, 1035–1038.

A.N. Varchenko, *Asymptotic Hodge structure on vanishing cohomology.* Izv. Akad. Nauk SSSR Ser. Mat. 45 (1981), no. 3, 540–591, 688.

J.C. Várilly, J.M. Gracia-Bondía, *Connes' noncommutative differential geometry and the standard model.* J. Geom. Phys. 12 (1993), no. 4, 223–301.

V. Voevodsky, *Triangulated categories of motives over a field.* in "Cycles, transfer and motivic homology theories", pp.188–238, Annals of Mathematics Studies, Vol.143, Princeton University Press, 2000.

W.C. Waterhouse, *Introduction to affine group schemes.* Graduate Texts in Mathematics, 66. Springer, 1979.

A. Zee, *Quantum field theory in a nutshell*, Princeton University Press, 2003.

W. Zimmermann, *Convergence of Bogoliubov's method of renormalization in momentum space*, Comm. Math. Phys. 15, (1969), 208–234.

Index